Decision Space

In *Decision Space: Multidimensional Utility Analysis*, Paul Weirich increases the power and versatility of utility analysis and in the process advances decision theory. Combining traditional and novel methods of option evaluation into one systematic method of analysis, multidimensional utility analysis is a valuable new tool. It provides new formulations of important decision principles, such as the principle to maximize expected utility; enriches decision theory in solving recalcitrant decision problems; and provides in particular for the cases in which an expert must make a decision for a group of people. The multiple dimensions of this analysis create a decision space broad enough to accommodate all factors affecting an option's utility.

The book will be of interest to advanced students and professionals working in the subject of decision theory, as well as to economists and other social scientists.

Paul Weirich is the Chair of the Department of Philosophy at the University of Missouri–Columbia and author of *Equilibrium and Rationality: Game Theory Revised by Decision Rules* (1998).

T0276219

Cambridge Studies in Probability, Induction, and Decision Theory

Decision Space

Multidimensional Utility Analysis

PAUL WEIRICH

University of Missouri–Columbia

CAMBRIDGE UNIVERSITY PRESS
Cambridge, New York, Melbourne, Madrid, Cape Town, Singapore, São Paulo

Cambridge University Press
The Edinburgh Building, Cambridge CB2 8RU, UK

Published in the United States of America by Cambridge University Press, New York

www.cambridge.org
Information on this title: www.cambridge.org/9780521800099

First published 2001
This digitally printed version 2007

A catalogue record for this publication is available from the British Library

Library of Congress Cataloguing in Publication data

Weirich, Paul, 1946–
 Decision space : multidimensional utility analysis / Paul Weirich.
  p. cm. – (Cambridge studies in probability, induction, and decision theory)
 Includes bibliographical references and index.
 ISBN 0-521-80009-9
 1. Decision making. 2. Utility analysis. I. Title. II. Series.
QA279.4 .W45 2001
519.5'35 – dc21   00-067494

ISBN 978-0-521-80009-9 hardback
ISBN 978-0-521-03803-4 paperback

For my daughters, Corinne and Sonia

Contents

Figures

Preface

The utility of an event depends on various factors. According to tradition, it may be broken down into the event's utility for each of the people it affects, or the utilities of the possible worlds it may yield, or the utilities of the goods and bads it realizes. Each form of utility analysis introduces a dimension along which an event's utility is dispersed. Utility analysis gains power and versatility if traditional methods of analysis are systemized to form one multidimensional method of analysis. My aim is to bring about this unification and thereby advance decision theory, which relies on analyses of an option's utility.

Multidimensional utility analysis both improves and enriches decision theory. It improves decision theory by providing new formulations of important decision principles, such as the principle to maximize expected utility. It also enriches decision theory by adding the power to solve recalcitrant decision problems. In particular, the richer theory handles cases in which an expert makes a decision for a group of people. The multiple dimensions of analysis create a decision space broad enough to accommodate all factors affecting an option's utility.

My theory's cornerstone is a new form of intrinsic utility analysis. It uses basic parcels of utility to which all other utilities are reducible. Other forms of utility analysis use utilities constructed from these basic parcels. Creation of a common framework for utility analysis ensures the accuracy and consistency of my versions of traditional forms of utility analysis. It shows how various types of utility are related. This framework also adds explanatory power and generates novel extensions of utility analysis.

The theory is systematic about idealizations. It itemizes and classifies its idealizations and thereby provides directions for eventually removing them to achieve greater realism. Although the theory is reliant on idealizations, it has practical import. It selects idealizations so that its principles guide the decision maker in the way that the idealized laws of science guide the engineer.

I advance decision and utility principles that involve concepts eschewed by operationists. For example, my principles treat degrees of belief and desire as primitive theoretical quantities (undefined, but not unexplained). They also introduce basic intrinsic desires, interpersonal utilities, and quantitative social power. These concepts lack the observational clarity the operationist prizes. But they compensate by making decision theory more systematic, powerful, and explanatory. My enterprise sacrifices the operationist's goals for the sake of the theoretician's goals. Its method suits the philosophical decision theory it yields. Philosophical decision theory is normative; that is, it investigates the way we ought to make decisions to be rational, rather than the way we make decisions. Although the empirical sciences may need operationist methods for their objectives, other methods better serve the objectives of philosophical decision theory.

This book is written for philosophers, economists, and behavioral scientists interested in decision theory. It is self-contained and presumes only high school algebra. It contributes new theoretical ideas but is accessible. Points for the specialist and not essential to the main argument appear in appendices to the chapters and to the book.

This is the first presentation of my version of intrinsic utility analysis and its systematization of traditional forms of utility analysis. My treatment of expected and group utility analysis occasionally draws on my prior publications (chapter notes specify sources). Many people influenced my project, which began in 1987. I would like to thank especially Wayne Davis, Ellery Eells, Richard Feldman, Alan Fuchs, Wlodzimierz Rabinowicz, J. Piers Rawling, Reed Richter, J. Howard Sobel, Peter Vallentyne, and two anonymous readers. I would also like to thank the National Science Foundation for a Summer Fellowship in 1988, the University of Missouri–Columbia Research Council for support during the summers of 1989 and 1991 and the academic year 1994–5, and the University of Missouri Research Board for support during the academic year 1994–5. At Cambridge University Press Terry Moore, Gwen Seznec, Robyn Wainner, and Matthew Lord were helpful at every step. My production editor, Brian MacDonald, artfully turned my manuscript into a book. They all have my lasting gratitude.

1

Objectives and Methods

The Occupational Safety and Health Administration (OSHA) regulates the workplace to reduce hazards to employees. Employers often challenge its decisions. Debate draws attention to many pressing issues in decision theory. To cover regulatory decisions made by government agencies for the public, decision theory must expand and reorganize. It needs an account of a group's probability and utility assignments. It also needs an account of decision making for others given uncertainty.

My project is the formulation of rules for rational decision making – that is, the construction of a normative decision theory. I seek rules powerful enough to handle the complexities of decisions made for another person or group of persons. In the cases targeted, a professional or group of professionals has some expert information and wants to use it to serve a client's goals. I expand and improve decision theory so that it offers practical guidance in these decision problems.

I assume that a rational decision maker deciding for herself adopts an option of maximum utility and then argue for various ways of calculating an option's utility. Decision theory gains range and depth by analyzing utility according to a multidimensional method that Section 1.1 explains and later chapters elaborate. Briefly, multidimensional utility analysis unifies methods of analyzing utility. It uses one method to explain the others and establish their consistency. As the basic method of analysis, I adopt a new version of a traditional method, intrinsic utility analysis. Following several chapters presenting methods of utility analysis, Chapter 7 applies multidimensional utility analysis to decisions made for others.

1.1. MULTIDIMENSIONAL UTILITY ANALYSIS

An agent's assignment of utilities to acts records her assessment of their instrumental value with respect to her ends. Utility analysis reduces an act's utility to other utilities generating the act's utility. Three traditional forms of utility analysis are group utility analysis, intrinsic utility analysis, and expected utility analysis. Group utility analysis, inspired by utilitarian moral theory, reduces an act's utility for a group to its utilities for the group's members. Intrinsic utility analysis, inspired by value theory, reduces an act's intrinsic utility for a person to the intrinsic utilities of realizing intrinsic desires and aversions. Expected utility analysis, inspired by probability theory, reduces an act's utility for a person to utilities of subjective chances for the act's possible outcomes. There are other traditional forms of utility analysis, but I concentrate on these three.

Multidimensional utility analysis provides a way of using the three methods of utility analysis together. It is fruitful because combining the methods suggests improvements of each, and new hybrid forms of utility analysis. It is explanatory because the methods' combination exposes the common source of their justification. It is versatile because for a wide variety of decision problems it offers an appropriate analysis of options' utilities. And it is powerful because it has the resources to address decisions made for others.

1.1.1. Conjoint Utility Analyses

Intrinsic utility analysis, expected utility analysis, and group utility analysis are generally used separately and then yield unidimensional utility analyses. Each's application analyzes a utility along a single dimension, or with respect to a single type of consideration. These forms of utility analysis may, however, be used conjointly and then yield multidimensional utility analyses. A conjoint application analyzes a utility along multiple dimensions, or with respect to multiple types of consideration.

Suppose a new safety standard is implemented. Under some simplifying assumptions, the act's utility for a group of three people – A, B, and C – may be divided into the act's utility for each person, as in Figure 1.1. The act's group utility may also be divided according to the act's realization of the group's basic intrinsic desires and aversions – say, desires for health, security, and prosperity – as in Figure 1.2. Each of these analyses is unidimensional. Combining the two analyses generates the analysis Figure 1.3 depicts. The conjoint utility analysis may be viewed as an analysis along a single dimension of considerations combining a person and an intrinsic desire. But because utility analy-

2

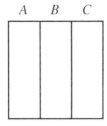

Figure 1.1. Group Utility Analyzed Person by Person

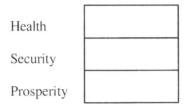

Figure 1.2. Group Utility Analyzed Goal by Goal

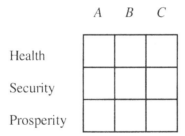

Figure 1.3. Group Utility Analyzed according to People and Goals

ses according to people and according to intrinsic desires and aversions are independently viable, I classify the conjoint analysis as a multidimensional utility analysis.

Two methods of analysis operate along two dimensions just in case they are independent ways of dividing considerations. A multidimensional utility analysis applies simultaneously two or more independent forms of utility analysis to obtain an option's utility. My utility theory combines intrinsic, expected, and group utility analyses. It is multidimensional because it elaborates three independent forms of analysis so that they are usable conjointly to obtain an option's utility. It has

3

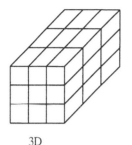 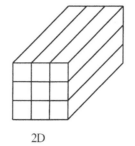 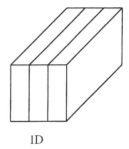

3D   2D   1D

Figure 1.4. Weight Spatially Analyzed

this status even though its forms of analysis are reducible to analysis in terms of certain fine-grained considerations.

The term multidimensional utility analysis arises from an analogy between utility analysis and forms of analysis involving spatial dimensions. Suppose, for instance, that we are analyzing the weight of a block of wood. Figure 1.4 depicts three possibilities. With respect to a three-dimensional coordinate system, we may obtain the block's weight by adding up the weights of its parts taken as disjoint contiguous unit blocks. A two-dimensional analysis obtains its weight by adding up the weights of its parts taken as disjoint contiguous bars with unit square cross sections. A one-dimensional analysis obtains its weight by adding up the weights of its parts taken as disjoint contiguous slabs one unit thick. The dimensions provide locations for assignments of parts' weights, added to obtain the whole block's weight.[1]

The dimensions of utility analysis are dimensions of utilities, or dimensions for spreading out factors contributing to utilities. The dimensions of a utility analyzed are the types of consideration contributing to the utility. One type is utilities for people. Another is utilities from realizations of intrinsic desires and aversions – for short, realizations of goals. Another is utilities of possible outcomes. The types of consideration form a space of reasons explaining the utility's value. In a decision problem, where an option's utility is analyzed, I call the space of reasons a *decision space*. The points of the space are locations for reasons tallied to obtain an option's utility.[2]

---

[1] To apply these methods of analysis to the weight of an irregular object, we may use the weights of the object's parts within unit blocks, bars, or slabs.

[2] The points of a decision space for an option are locations of reasons for or against the option. The reasons themselves may be utilities or components of utilities. For

4

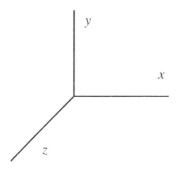

Figure 1.5. Decision Space

The opposite of spreading reasons along a dimension is agglomerating those reasons. Considerations along a dimension may be agglomerated to reduce the dimensionality of a decision space. For example, the factors making up the utility of a governmental policy for a person may be agglomerated to obtain the policy's utility for the person. Doing this for all people reduces a decision space with dimensions for people and for goals to a decision space with a single dimension for people.

Imagine the three-dimensional decision space that Figure 1.5 represents. The utility of an xyz-point is the finest unit of analysis. The utility of a line parallel to the z-axis may condense on the point where it intersects the xy-plane. The utility of a plane parallel to the yz-plane may condense on a point where the x-axis intersects it. Because the dimensions represent types of consideration, these maneuvers reduce one or two types of consideration to another. They make the space of reasons less fine-grained.

Utility analysis reduces an option's utility to reasons for or against the option, or utilities attached to those reasons. The reduction may be more or less fine-grained. A one-dimensional reduction is less fine-grained than a two-dimensional reduction. Moreover, some reductions

example, along the dimension of people a point is a person. A reason located at a person may be a utility for the person, or the realization of a desire of the person. The nature of the reasons depends on whether they are to be directly added to obtain the option's utility, or first weighted before adding to obtain the option's utility. The main task of reasons located along a dimension is to separate pros and cons of a certain type as a preliminary for an analysis of the option's utility. Although I prefer treating reasons as utilities, other types of reasons may also occupy a role in utility analysis.

decrease resolution by using intervals along a dimension rather than points. That is, they use reasons located at sets of points in a dimension rather than at single points in the dimension. For example, along the dimension of people, the reasons may concern groups rather than individuals.

Some utility analyses increase resolution, not by adding a dimension, but by moving from intervals to points along the same dimension. In Figure 1.5 the utility of an interval along the x-axis may be analyzed in terms of the utilities of the points it contains. This analysis works with a single dimension, or considerations of a single type. The considerations supporting the points' utilities combine to support the interval's utility.[3]

How many dimensions does utility analysis need? The answer depends on the task to perform. Consider the computation of utility for a group of people. I assume that given idealizations discussed later the source of utility is the realization of goals. So an analysis's task is to compute a group's utility assignment from this source. My analysis locates a goal's realization at the goal realized. Thus, it needs the dimension of goals. If the group considered comprises all people, group utility may be parceled out along the dimension of goals without attending to whose goals they are. But normally an analysis considers a smaller group and confines itself to goals the group's members have. It then locates a goal's realization at a person and at a goal the person has. Therefore it also needs the dimension of people. If certainty were universal, the dimensions of goals and people would suffice. To compute group utility given uncertainty, however, an analysis must take into account possible realizations of goals. To count these possibilities, my analysis considers possible worlds in which the goals are realized. It locates realizations of goals at possible worlds. Possible worlds thus form a third dimension of utility analysis. My

---

[3] Although coarse-grained considerations of one type may reduce to fine-grained considerations of the same type, and considerations of one type may reduce to finer-grained considerations of another type, these reductions do not eliminate the considerations or types of consideration reduced. A nonbasic consideration or type of consideration still has a role in utility analysis. In particular, a nonbasic dimension still constitutes a type of consideration with respect to which a utility may be analyzed, a type of consideration distinct from the type of consideration to which it is reducible. Such noneliminative reduction certainly takes place in analysis using spatial dimensions. The dimensions of spatial analysis are reducible to various sets of spatial points, but still are distinct and ground independent methods of spatial analysis.

plans for utility analysis therefore require three dimensions: goals, people, and worlds.

To compute a person's assignment of utility to an option given uncertainty, an analysis needs only two dimensions: goals and worlds. It may suppress the dimension of people by considering only goals the person has. A person's utility assignment depends on the realization of his goals only. Other people's goals matter only insofar as they affect his goals.

Coarse-grained versions of the three dimensions targeted are sometimes useful. To calculate the utility of a policy for a large group from its utility for subgroups, an analysis may make the dimension of people more coarse-grained to form a derivative dimension of groups. Similarly, it may replace the dimension of worlds with a more coarse-grained derivative whose units are sets of worlds, or states.

Dimensions besides the three mentioned are also useful for certain tasks, such as deciding between options according to something less encompassing than their outcomes' utilities. Suppose we want a type of utility that focuses on an option's causal consequences rather than on its entire outcome. Then we may locate realizations of goals at basic events and calculate utility in a set of consequences according to the goals realized within it. This procedure introduces a causal classification of events and so a causal dimension of utility analysis. Alternatively, we may want a type of utility that focuses on an option's aftermath rather than on its entire outcome. Then we may locate realizations of goals at times and calculate utility during an interval according to the goals realized within it. This procedure introduces a temporal dimension of utility analysis. Still other possibilities introduce spatial or spatiotemporal dimensions of utility analysis. For example, to compute the utility of a weather system such as a hurricane, one might tally its consequences on various days in various areas. Similarly, to obtain the utility of a donation to charity, one might tally its consequences at various times for the various people affected. My multidimensional framework may expand to include these additional dimensions of utility analysis, but for simplicity it focuses on the dimensions of goals, people, and worlds.[4]

---

[4] Of course, besides adding dimensions of utility analysis, one may alter the framework for utility analysis in other ways. For instance, one may use analysis of the utility of a commodity bundle or a dated commodity bundle – rather than the utility of a proposition – as the foundation of multidimensional utility analysis. Also, prior

Multidimensional utility analysis differs from multiattribute utility analysis. Attributes are factors affecting a utility's value. Perhaps a car's utility comes from attributes such as performance, reliability, economy, and styling. All the attributes belong to a single dimension, namely, the type of consideration pertaining to a driver's goals. Although the analysis of the car's utility involves multiple attributes, it uses only a single dimension. In contrast, a multidimensional utility analysis involves considerations falling into two or more categories, each of which suffices for a complete, independent utility analysis. Furthermore, a utility's multiattribute analysis may not divide the utility into components whose utilities add up to the utility analyzed. Two attributes may complement each other so that their combination's utility is greater than the sum of their utilities. For instance, the utility of a shopping basket with bacon and eggs may be greater than the sum of the utilities of the items it contains. In contrast, a utility's analysis along a dimension divides the utility into components that add up to the utility analyzed.

Let me also distinguish my type of multidimensional utility analysis from another type of analysis that goes by the same name. Some theorists debate whether utility should be measured on a single scale or on distinct scales assigned to incomparable sources of utility, such as pleasure and fulfillment of moral obligations. Etzioni (1986) and Brennan (1989) propose to handle incomparability by introducing multiple utility scales they call dimensions. My version of multidimensional utility analysis does not introduce dimensions to accommodate incomparability. It analyzes a single type of utility in diverse ways. The various methods of utility analysis are not motivated by incomparability but in fact assume the comparability of the objects to which utilities are assigned. This assumption is part of an idealization, explained later, that decision makers have quantitative utility assignments. I take it as understood that their assignments are justifiably quantitative with respect to a single scale. I acknowledge the problem of incomparability, presented, for example, in Chang (1997). My theory simply skirts the problem. It does not assume that utilities are comparable in all cases. It assumes that utilities are comparable in some cases, and treats only those cases.

Multidimensional utility analysis, as I present it, offers practical assistance with decisions made for others. These decision problems

to a utility analysis, one may use sequential decision analysis to separate a decision's consequences into its consequences for future decisions and its other consequences.

may be soluble theoretically with unidimensional utility analyses that decompose an act's utility for a client along a single dimension. But if the client is a group of people with diverse goals, and the act's outcome is uncertain, it is convenient to decompose the utility along multiple dimensions. Decisions made for groups when risk is an issue profit from the three methods of utility analysis. Expected utility analysis handles the uncertainty involved. Intrinsic utility analysis handles attitudes toward risk. Group utility analysis handles conflicts between the group's members. As Chapter 7 demonstrates, the combination of expected, intrinsic, and group utility analyses is a powerful mechanism for making decisions for others.

1.1.2. Precedents for Multidimensional Utility Analysis

My form of multidimensional utility analysis advances independent forms of utility analysis applicable conjointly. Some authors present theories with the same purpose. Keeney and Raiffa (1976) present a method of utility analysis that is multidimensional. Their method is a predecessor of mine, although it has an operationist orientation mine lacks. They consider situations in which a decision maker has multiple objectives, for example, situations in which a job applicant seeks both income and job satisfaction. Each objective provides a scale along which an option is evaluated. Utilities along these scales combine to form the option's utility. The scales are not dimensions along which utility is located but rather devices by which utility is measured. The various objectives generating the scales form a single dimension along which an option's utility is dispersed – roughly, the dimension of goals. When an option's outcome is uncertain, however, expected utility analysis is used in tandem with an analysis according to multiple objectives. This simultaneous use of two independent forms of analysis is multidimensional. It spreads an option's utility across objectives and also across possible outcomes. It is a combination of intrinsic and expected utility analyses.[5]

---

[5] Mullen and Roth (1991: chap. 7) also present a combination of expected utility analysis and intrinsic utility analysis that they call multiattribute utility theory. Kleindorfer, Kunreuther, and Schoemaker (1993: sec. 4.2.3 and app. C) discuss a similar combination of expected utility analysis and intrinsic utility analysis that they call multiattribute utility analysis. Both works take the same general approach as Keeney and Raiffa (1976).

The main structural difference between Keeney and Raiffa's and my multidimensional form of utility analysis concerns group utility analysis. Keeney and Raiffa omit it, largely because their operationism does not condone the interpersonal utilities that group utility analysis requires. To treat decisions involving groups, they introduce a social planner who makes decisions for a group and whose objective is to serve each member of the group (1976: chap. 10). They use the social planner's estimate of an act's utilities for the group's members as part of an intrinsic utility analysis of the act's utility for the social planner. They do not use the act's utilities for the members as part of a group utility analysis. My form of multidimensional utility analysis runs deeper than theirs because it provides a justification of the preferences of a utilitarian social planner.

Two methodological differences are also important. First, Keeney and Raiffa do not attempt to systematize their methods of utility analysis by deriving them from a single principle. There is nothing in their work corresponding to my derivation of intrinsic, expected, and group utility analyses from the principle of pros and cons (see Section 1.2.1). Second, they define probability and utility operationistically, which limits the explanatory value of these concepts (see Section 1.3.2).

Keeney and Raiffa's method of multidimensional utility analysis is finely crafted and a source of inspiration for my own method. I enrich their method, make it more systematic, and enhance its justificatory power by taking a nonoperationistic approach to multidimensional utility analysis.

Broome (1991) provides another precedent for my theory. He combines expected, temporal, and group utility analyses. My form of multidimensional utility analysis differs structurally. It includes intrinsic utility analysis. This is crucial. On my view, some dimensions are more basic than others. One dimension is more basic than another if the reasons located along it are more basic than the reasons located along the other dimension, that is, if the first set of reasons explains the second. Because realizations of goals are the source of utility, the dimension of goals is more basic than other dimensions. Realizations of goals explain utilities located at people and worlds. They also explain utilities assigned to times. As a result, intrinsic utility analysis explains group, expected, and temporal utility analyses. My inclusion of intrinsic utility analysis, and the underlying basic dimension of goals, makes my form of multidimensional utility analysis more explanatory and systematic. It treats chances of realizations of goals as basic reasons for utilities and other reasons as derivative.

Although Broome investigates temporal utility analysis, he expresses doubts about both intrinsic and temporal utility analyses because he has misgivings about the separability of utilities along the dimensions of goals and time – that is, the possibility of allocating the parts of an option's utility to goals and times without omission or double counting (1991: 25, 29). Chapter 2 argues for the separability of utilities along the dimension of goals. Their separability along this dimension justifies their separability along the temporal dimension, too. Intrinsic utility analysis, the foundation of my multidimensional form of utility analysis, justifies temporal utility analysis. Although, for simplicity, I put aside temporal utility analysis, the type of utility analysis I add to Broome's assortment supports it.

Broome's theory and mine also differ methodologically. I make desire rather than goodness the basis of utility. Then to give desire a structure like goodness's, I rely on idealizations such as the rationality of agents. Consequently, idealizations are more prominent in my theory than in Broome's. Also, I make greater departures from operationism. Although Broome is willing to entertain interpersonal utility, he defines probability, personal utility, and interpersonal utility operationally.[6] This makes basic principles of utility analysis truths of definition rather than tools of explanation (see Section 1.4.2). It also makes otiose any attempt to justify the various forms of utility analysis with a general principle of rationality such as the principle of pros and cons (see Section 1.2.1).

Although Keeney and Raiffa (1976) and Broome (1991) provide important precedents for multidimensional utility analysis, my non-operationistic theory has significant advantages. It formulates intrinsic, expected, and group utility analyses in a way that provides deeper, more systematic explanations of rational decisions. It also provides more accurate versions of these forms of utility analysis and makes them versatile enough to handle the complexities of decisions made

---

[6] Instead of arguing for the separability of certain considerations, Broome (1991) assumes their separability in connection with the representations theorems his operational definitions invoke. See, for example, his chap. 4, esp. p. 70, and chap. 10, esp. p. 202, where separability assumptions allow additive utility functions to represent the betterness relation among composite goods. I take utility as primitive and argue for certain types of separability to justify utility assignments instead of assuming those types of separability to define utility assignments.

for others. My reduction of the various forms of analysis to a single form of analysis, intrinsic utility analysis given uncertainty, achieves this systematization and greater accuracy.

1.2. VERIFICATION, JUSTIFICATION, AND VINDICATION

As later chapters advance the three main methods of utility analysis, I review their credentials. First, I verify their accuracy using intuitions about cases and principles. Second, I justify them by deriving them from a basic principle of utility analysis that Section 1.2.1 presents, the principle of pros and cons. Third, I vindicate them by showing that utility, as they analyze it, reliably indicates rational decisions. Sections 1.2.1 and 1.2.2 explain the type of justification and vindication sought.

1.2.1. The Principle of Pros and Cons

Multidimensional utility analysis is a promising tool for complex decisions such as decisions made for others. For it to succeed, however, its three independent forms of utility analysis must be consistent when applied conjointly. Consistency may require some mutual adjustment. My method of ensuring consistency is to derive each of the three forms of utility analysis from a single general principle, the principle of pros and cons. This principle of utility analysis is old and familiar.[7] It says to list an option's pros and cons and to attach utilities to them, positive or negative according as they are pros or cons. The utilities serve as weights indicating the importance of considerations. Then one adds the pros and cons using their utilities to obtain the option's utility. The principle of pros and cons is simple and fundamental, but clear and precise applications are difficult. One needs directions for listing pros and cons and attaching weights to them. Applications of the principle in later chapters provide these directions. They show that intrinsic, expected, and group utility analyses are all methods of adding pros and cons.

The first task in applying the principle of pros and cons is to separate considerations regarding an option into pros and cons. The separation of considerations must satisfy two conditions: (1) no consideration may be counted twice, and (2) no relevant consideration

---

[7] Benjamin Franklin (1945: 280–1) advances a version of the principle of pros and cons.

may be omitted. Adding the utilities of the considerations regarding an option does not yield the option's utility unless these conditions are satisfied. For no consideration should influence the option's utility twice, and no relevant consideration should fail to influence the option's utility altogether.

Dividing relevant considerations so that none is omitted or double-counted is difficult. Many considerations are mixed in everyday deliberations. In deciding between jobs available, one might try to separate income and job satisfaction. But this does not work since income itself is a source of job satisfaction. Evidently, the goals of no omission and no double-counting push in different directions. The more considerations entertained, the less likely is omission, but the more likely is double-counting.[8]

The second step in applying the principle of pros and cons is to obtain the utilities of considerations. A difficulty in this second step is making assessments of utility quantitative. In some cases one consideration clearly has more utility than another, but not clearly a certain number of times more utility. Because of this difficulty, the principle of pros and cons seems impractical. Like most quantitative methods of decision making, it appears unrealistic. Quantitative methods may work for insurance companies seeking to maximize expected profits in the light of extensive statistical data, but they seem ungrounded in other contexts.

To put aside this objection, I idealize. I assume circumstances justifying the quantitative aspects of applications of the principle of pros and cons. The quantities assumed may be taken as ideal

---

[8] Separating considerations into pros and cons is difficult because the concurrence of objects of desire may have a utility different from a combination of the utilities of the objects occurring in isolation. Such complementarity thwarts methods of obtaining the utility of a composite as a function of the utility of its components. Suppose that u_1 and u_2 are utilities of objects and u is a function giving utilities of composites. A function must have the same value in cases where its arguments are the same. So if $u(u_1, u_2)$ has a certain value for certain values of u_1 and u_2, then it must have the same value whenever u_1 and u_2 have those values, whatever the objects that generate those values. For example, suppose that the utility of a left shoe is 1 and the utility of a left glove is 1. Then their joint utility is $u(1, 1)$. Also suppose that the utility of a right shoe is 1 and the utility of a right glove is 1. Then their joint utility is also $u(1, 1)$. It follows that the utility of a left glove and a right glove is $u(1, 1)$, and that the utility of a left shoe and a right shoe is $u(1, 1)$. This consequence ignores the utility gain from trading unmatched for matched pairs of gloves and shoes. The mistake arises no matter how the function u combines utilities.

entities. The assumption of quantitative probabilities and utilities is an idealization rather than a mere restriction because the assumption puts aside explanatory complications concerning imprecision (see Section 1.3.1). Applications of the principle of pros and cons under idealizations are still useful. They serve as instructive guides in realistic decision situations, especially ones that approximate the ideal case.

Also, although the principles of utility analysis I advance rest on an idealization, the principles follow from a completely general version of the principle of pros and cons. The idealization serves only to make the principle of pros and cons applicable, not to limit its application. The principles of utility analysis apply wherever the quantitative input for them exists. Also, the principles of utility analysis for options are the same as the principles of utility analysis for outcomes of options and for options conditional on states of the world, and for various types of utility such as group, temporal, and causal utility. This generality explains subtleties of the principles; their formulations for the paradigm cases are adjusted so that they generalize. Besides increasing the scope of the theory, generalization of the principles makes the theory more explanatory. For it would be hard to maintain that the principles explain the rationality of decisions if they included restrictions seeking only to put aside recalcitrant cases.

To justify the three forms of utility analysis using the principle of pros and cons, I investigate each form's method of separating considerations and assigning utilities to them. First, I show that each form uses a method of separating considerations that neither omits nor double-counts any relevant consideration. Then I show that considerations are assigned utilities in a suitable way. Deriving the three forms of utility analysis from the single principle of pros and cons, using appropriate principles of separation into pros and cons for each kind of utility analysis, produces a systematic theory of utility analysis. All the methods of utility analysis derive from the principle of pros and cons and so agree with each other.

The derivation of all forms of utility analysis from the principle of pros and cons establishes their correctness and, a fortiori, their consistency. Nonetheless, for vividness, the appendix demonstrates that the varieties of utility analysis work together in a systematic way to explain why certain decisions are rational and others are irrational. To show that the three methods of utility analysis form a systematic theory, the appendix investigates various combinations applied to decision problems. For example, it shows that the utility of a possible outcome appearing in an expected utility analysis may be computed using intrinsic utility analysis. Also, it shows that given certain restrictions, when expected utility analysis is applied to acts and group utility analysis is used to obtain the utilities of outcomes, the result is the same as when group utility analysis is applied to acts and expected utility analysis is used to obtain the utilities of the act for group members.

My argument for the three forms of utility analysis derives them from the principle of pros and cons and claims that the derivation justifies them. Let me say a little more about the justification provided. I do not advance a theory of justification. The literature on justification in epistemology, for example, warns of the complex issues such a theory must resolve. But I do clarify justification as I understand it.

Justification goes beyond verification. For example, justifying a decision is explaining its rationality, not just verifying its rationality. When a decision's rationality is in doubt, one may remove the doubt without explaining the rationality of the decision, hence without justifying the decision. One may provide only verification that the decision is rational. This happens when the decision is shown to be required by coherence with other rational decisions. Its justification, in contrast, goes beyond such coherence. It involves the decision's promotion of the agent's goals, provided they are rational. Verifying a preference's rationality and justifying the preference differ in a similar way. One may demonstrate a preference's rationality without explaining its rationality. A preference's coherence with other preferences may verify that the preference is rational if the others are, but does not explain the rationality of the preference. The explanation of its rationality involves the features of the objects it compares.

The distinction between verification and justification applies in a similar way to principles of utility analysis. Verification of a principle shows its correctness. Justification of a principle explains its correctness. Testing and checking a principle differ from explaining it. These endeavors, if successful, support the principle. But we may see that a principle is correct without knowing why it is. Explanation goes beyond support. Hence verification of a principle may stop short of its justification. Checking against intuitions furnishes support for a principle's correctness. Derivation from the principle of pros and cons furnishes an explanation of its correctness, and so a justification.

My goal is to justify principles of utility analysis. For each principle of utility analysis advanced, I seek an explanation of the rationality of conformity with the principle. The explanation proceeds by derivation from the principle of pros and cons, the foundation of utility analysis. The claim that the principle of pros and cons justifies a form of utility analysis means that the principle (if we grant ancillary assumptions about the relevant pros and cons and their weights) yields the form of utility analysis.

The claim that the principle of pros and cons provides a utility analysis's justification singles out the most salient feature of the analysis's justification. In other cases of justification, the most salient feature may not be a principle but the input for it. For instance, we may say that a person's preferences justify his decision. Here the complete justification involves the decision principle to adopt a top option, as well as the person's preferences. But the preferences are the most salient factor. In typical cases, I use the principle of pros and cons to justify a utility analysis, a utility analysis to justify an option's utility, two options' utilities to justify a preference between the options, and preferences between options to justify a decision. The justifications are explanatory and work by derivation. Some identify a salient principle and others identify salient input for a principle.

Note that although I claim that options' utilities may justify preferences among the options, that is, explain their rationality, I do not claim that preferences always have this type of justification. Some preferences are basic and explain options' utilities. The type of justification, being a species of explanation, may vary from case to case. Under the idealizations, an agent's goals and the utility assignments they generate provide an argument for preferences capable of justifying them. But whether that argument actually justifies the preferences depends on the agent's circumstances.

My justification of utility principles appeals to normative principles. I derive principles of utility analysis from the more basic normative principle of pros and cons. Some standards of rationality may have special, nonnormative justifications. Velleman (1997) imagines a type of justification for basic normative principles that is not normative. He argues that normative principles of decision have a theoretical grounding in the nature of action and its constitutive aim. The grounding appeals to theoretical rather than practical reason. Such theoretical grounding may be possible for decision rules and utility principles, but I do not attempt it. The justification I provide is normative and grounded in practical reason.

1.2.2. Utilities and Decisions

Throughout this book, I assume the principle of pros and cons. I also assume the decision rule to adopt an option of maximum utility, that is, to adopt an option whose utility is at least as great as the utility of any other option. The decision rule, which has widespread support, stems from the more basic rule to maximize informed utility. Maximizing informed utility ensures success in terms of one's objectives, but ignorance often impedes maximization of this quantity. To circumvent the problem, utility responds to information available, even if incomplete. It is the realistic counterpart of informed utility, and maximizing it is the realistic way of pursuing maximum informed utility.[9]

The decision rule to maximize utility creates a standard for evaluating decisions. It is not advanced as a decision procedure. I do not say that a decision maker must apply the rule to reach a decision. I formulate the standard as a rule in order to indicate that maximizing expected utility is a goal of rational agents. Where obstacles impede its attainment, rational agents pursue the goal in ways reasonable, given those obstacles and their other goals. They may, for example, cultivate character traits and make decision preparations that promote utility maximization.

My utility principles and decision rules are designed for ideal cases. They provide a means of calculating an option's utility and using it to guide decisions. I formulate a decision theory by filling out the framework they provide. However, the entire decision theory, the whole package of decision rules and utility principles, wins support, beyond its native plausibility, from its agreement with the verdicts of intuition about rational decisions.[10] The support a principle of

---

[9] I take maximizing informed utility as the primary goal of decision, and maximizing utility, and maximizing expected utility, as secondary goals for cases where ignorance creates obstacles to the primary goal. Beebee and Papineau (1997: 238–43) argue that maximizing expected utility is the primary goal of rational decision. From it they derive the goal of gathering information. However, the primacy of the goal to maximize informed utility best explains why a rational but unsuccessful decision evokes rational regret.

[10] Some theorists, for example, Gauthier (1986), Bratman (1987: 23), Slote (1989), and McClennen (1990, 1992), dispute the decision rule to maximize utility. Evaluating their claims goes beyond this book's scope. I work within the orthodox school that accepts the decision rule to maximize utility. For some recent literature on the issues, see Rabinowicz (1995), Skyrms (1996: 38–42), Velleman (1997), and Coleman and Morris (1998).

utility analysis derives from the success of the decision theory to which it belongs I call *vindication*.[11] Justification is internal to utility theory, whereas vindication operates in the wider realm of decision theory. Justification explains the rationality of complying with a utility principle, whereas vindication supports the utility principle by showing that conforming utilities yield rational decisions. Vindication of utility analyses shows that the decisions the analyses generate are rational according to our judgments about particular cases and general principles.

The vindication of any method of utility analysis, either uni- or multidimensional, therefore involves more than derivation from a fundamental principle of utility analysis, such as the principle of pros and cons. It requires establishing that the method's assignment of options' utilities is a reliable decision tool. This is why utility analysis is part of decision theory. Because our conception of utility is anchored in decision, the fundamental data supporting any form of utility analysis are judgments about the rationality of decisions. In ideal conditions, every form of utility analysis must yield rational decisions when joined with the principle to choose an option with maximum utility. The forms of utility analysis the principle of pros and cons generates meet this standard. If considerations regarding an option are separated and assigned utilities properly, the sum of their utilities yields a utility for the option that indicates its choiceworthiness given appropriate idealizations.

My vindication of principles of utility analysis by their application to decision problems takes utility principles and decision rules to have independent normative force. This independence assumes that utility is a primitive theoretical entity rather than by definition a quantity maximized by rational decisions. The decision rule to maximize utility has no normative force at all if utility is defined in terms of rational decisions. However, taking utility as primitive, a stance Section 1.4 defends, the decision rule has normative force. Also, its normative force is distinct from, for instance, the force of the utility principle asserting that an option's utility equals its expected utility.

As I have characterized them, justification and vindication of a form of utility analysis differ in scope and resources. A justification uses basic

---

[11] Feigl (1950: 122) suggests the term vindication. He uses it for an appeal to ends. The ends I have in mind are theoretical.

principles in the theory of utility analysis to extend and elaborate the theory. Using the principle of pros and cons to derive a form of utility analysis justifies the analysis. A vindication, in contrast, supports a form of utility analysis by showing that the whole theory of utility analysis, encompassing the principle of pros and cons as well as the form of utility analysis, agrees with intuitions about rational choice. I vindicate expected utility analysis, for example, by conjoining it with the decision rule to maximize utility and checking the results against intuitions about rational choice. In general, to ensure a utility analysis's vindication, I specify it so that when my decision theory incorporates it, intuitions about rational choice support the whole.

To bring out the distinction between justification and vindication, consider extensions of the three basic forms of utility analysis. Those forms of analysis apply to many types of utility. They apply to temporal utility, for example. Their justification is independent of the type of utility to which they are applied. However, not every type of utility suits the decision rule to maximize utility. Although my methods of utility analysis apply to utility defined in terms of past or future goals, the decision rule requires utility defined in terms of current goals. So requiring vindication of a type of utility analysis – showing its success in decision theory – raises the bar. Not only does the utility analysis need the right form. It also needs the right content. It has to involve a type of utility suitable for directing decisions.

My justification of methods of utility analysis is more thorough than my vindication of them. For analyses of conditional utilities, in particular, I focus on verification and justification rather than vindication. I use intuitions about conditional utility, instead of intuitions about rational decisions, to support the analyses. However, the success of the whole decision theory advanced also provides some vindication of the analysis of conditional utilities.

1.3. IDEALIZATIONS

My decision rules and principles of utility analysis rely on idealizations. The idealizations they employ are standard but strong. The rules and principles are theoretically attractive and offer practical guidance in decision problems despite their strong idealizations. They succeed because their idealizations meet certain explanatory standards. This section presents a general account of idealizations and uses it to motivate my theory's idealizations.

1.3.1. The Purpose of Idealizations

Multidimensional utility analysis assumes that agents and their situations are ideal in various ways. Some theorists repudiate such idealizations. They say that rationality for ideal agents in ideal situations does not illuminate rationality for humans. For instance, Foley (1993: 158–62) rejects the common idealization that agents are "perfect calculators." He observes that a person's degree of rationality is not the degree to which she approximates the standards of rationality for ideal agents. A person may be fully rational despite falling short of those standards. This observation is accurate, but justifications for idealizations may appeal to roles besides the measurement of rationality.

The principal role of idealizations is to facilitate partial explanations of phenomena by putting aside some explanatory factors to highlight others. Idealizations sacrifice generality to display the workings of certain critical factors. The explanatory role of genuine idealizations distinguishes them from mere restrictions or boundary conditions. Restrictions also simplify the formulation of exact principles. But they need not isolate explanatory factors. They just eliminate factors difficult to handle. For example, a theory of motion that puts aside friction idealizes, whereas one that treats only objects rolling on rails restricts.

In decision theory idealizations isolate factors regulating utilities and rational decisions to show how those factors function. Principles of decision and utility analysis should avoid mere restrictions. If they do, they have greater generality and explanatory power, even if they are idealized rather than completely general and yield partial rather than complete explanations of a person's rationality.

Besides the partial explanations they promote, my idealizations have two additional justifications. The first is practical. Because idealizations yield partial explanations, they yield partial understanding. This partial understanding has practical value. It reduces the scope of factors not understood and gives a decision maker his bearings. A decision maker gains by understanding some if not all factors behind a decision's rationality. The second additional justification is theoretical. An idealized theory builds a framework for a more general theory. Later, one may rescind some idealizations while retaining the framework of the original theory, and thus transform its partial explanations into more complete explanations of rational decision making.

Idealizations simplify the formulation of exact principles. In formulating principles of utility analysis, I use idealizations since I aim for exactness. Other approaches to decision theory, especially in practically orientated disciplines such as management science, compromise exactness for the sake of principles that are readily applicable to realistic decision problems. These other approaches are useful and intellectually stimulating. But my objective is theory construction, the construction of an explanatory theory that promotes understanding, not practicality. It is better given this objective to obtain a principle that is exactly right in idealized cases than to obtain a principle that is approximately right in realistic cases. Precision promotes explanation and understanding. Precise principles for ideal cases generate partial if not complete explanations. They describe exactly the influence of some relevant factors. Approximate principles for realistic cases often lack explanatory value if not derived from idealized but exact principles, whose explanatory value they inherit. To explain rational decision making, it is therefore more promising to use idealizations to obtain precise principles than to forgo idealizations and settle for rough rules of thumb.

The choice between precision for ideal cases and approximation for realistic cases raises deep issues in philosophy of science about theory construction and explanation. I do not try to resolve these issues in a general way but merely address the case at hand, decision theory. I claim that the explanatory goals of normative decision theory are well served by precision in ideal cases, and hope that the decision theory I present demonstrates this.

Although decision principles based on idealizations have mainly a theoretical function, they do have practical uses. Knowing what is rational in an ideal case serves as a guide in realistic cases, especially ones that approximate the ideal case. For example, the idealization that probabilities and utilities are available in decision problems is unrealistic. Quantitative precision is elusive. But this does not mean that a decision theory for quantitative cases has no value in realistic decision problems. Despite recourse to idealizations, my theory places certain constraints on decisions in realistic cases and furnishes a useful structure for deliberations. It also provides instructive examples of rational decision making.

Furthermore, precise, ideal principles may govern some aspects of a realistic decision problem even if they do not govern all aspects of the problem. This partial precision may be helpful. To see how, consider some examples outside decision theory. In monitoring a checking account, it is helpful to be precise about the transactions one records, even if one has forgotten to record some transactions; uncertainty is then restricted to the unrecorded transactions. Similarly, in applications of statistics it is helpful to be precise after making assumptions that are only approximately satisfied, because uncertainty is then restricted to the areas covered by the assumptions. In the same way, exact decision principles assuming idealizations may treat some matters with precision even if the idealizations hold only approximately. The exact principles are helpful because they restrict imprecision to the areas covered by the idealizations; they prevent the spread of imprecision.

I do not fully explain how principles for ideal cases provide guidance in realistic cases. That is worth doing. But the explanation is bound to be difficult and complex and is not necessary here, because the use of ideal principles in realistic cases is familiar from the use of ideal laws of science in engineering problems.[12]

1.3.2. My Idealizations

My decision principles and methods of utility analysis make two large idealizations. Both are standard idealizations in decision theory, although they are generally in the background. The large idealizations are: (1) rational ideal decision makers and (2) ideal decision situations. Their job is to make conditions perfect for satisfying decision rules and utility principles. They put aside obstacles that complicate explanations of rational decision making. Let us consider each of these large idealizations in turn.

According to the idealization about decision makers, they have no cognitive limitations. They know all logical, mathematical, and other a priori truths.[13] They know their own minds perfectly and never wonder about their beliefs and desires. They consider all their options. They

---

[12] For a comparison of idealizations in physics and in economics, see Lind (1993). Hamminga and De Marchi (1994) offer a collection of articles on idealizations in economics.

[13] For simplicity, this feature of the idealization is stronger than necessary. It suffices if agents know all a priori truths useful to them for making decisions that meet the goals of rationality given uncertainty about a posteriori matters. Weakening this way alleviates doubts about realizability. Some may suspect that the conceptual prerequisites of a priori omniscience are unsatisfiable, access to concepts being limited even for cognitively unlimited agents.

think instantaneously and without effort. These idealizations about cognitive powers remove obstacles to the attainment of goals of rationality.

The idealization also includes the rationality of agents. Their rationality's scope depends on the context. When evaluating principles of utility analysis, I assume that agents have rational goals and beliefs, and never make irrational decisions. When evaluating decision rules, I assume that agents follow all rules of rationality for beliefs and desires and decide rationally, except perhaps in the current case. The objective is to ensure that the principle or rule evaluated has rational input and that its output is used rationally so that it does not have to adjust to irrationalities in its applications' contexts.[14] An ideal agent who complies with it, assuming its correctness, is then fully rational.[15]

The idealization about rationality puts aside cases in which a violation of one principle of rationality motivates a compensating violation of another principle of rationality. I assume that a decision maker is fully rational except perhaps for violating the principle currently under consideration. When arguing for a principle analyzing an option's utility, I assume that an agent is fully rational concerning the input for the analysis. Also, I assume that the agent is fully rational in using options' utilities to reach decisions. In other words, I assume that the agent is fully rational except perhaps in assigning utilities to options.

---

[14] It is rational for an ideal agent to decide according to his current utility function, given its full rationality. Its full rationality rules out, for example, pure time preference, that is, discounting future desires just because they are future desires. If a person is fully rational, he has higher-order desires for the satisfaction of future desires. These higher-order desires influence his current utility assignment. For a discussion of decision rules and pure time preference, see Weirich (1981), Parfit (1984: chap. 6, sec. 45), Fuchs (1985), and Schick (1997: 69–71). In Parfit, see especially the critical version of the present aim theory.

[15] When I assume that decision makers are inerrant, I do not assume that they are infallible. It is possible for them to decide irrationally, although, in fact, they never do. The fallibility of decision makers permits a straightforward introduction of the utility of an option, given an assumption that some irrational decision is made. Although the assumption may be counterfactual, utilities with respect to such counterfactual conditions are sometimes crucial for the justification of a decision.

Also, although I assume that thought is instantaneous, I separate the time of deliberation and the time of decision so that it does not follow that an agent knows his decision as he deliberates. However, if the times of deliberation and decision were merged, one may still separate the two processes according to the inferences grounding them. Utility comparisons of options made in deliberations rest on basic beliefs and desires. Decisions rest on utility comparisons of options. Knowledge of the decision reached may not be a premiss in the inferential process yielding utility comparison of options.

Then I argue that to be rational given the idealizations, options' utilities must equal their values according to the principle of utility analysis. On the other hand, when evaluating the decision principle to maximize utility, I assume fully rational utilities for options but do not assume a decision of maximum utility. To avoid begging crucial questions, the idealization of rationality varies in content as I shuttle between decision rules and principles of utility analysis.

The idealization about agents also assumes the formation and accessibility of the probabilities and utilities needed for decision rules and utility analysis, in particular, expected utility analysis. Imprecision complicates decision goals in nonquantitative cases because without quantitative summaries of salient factors those goals must attend to more wide-ranging sets of considerations. Furthermore, agents are assumed to have stable goals, goals that are constant given suppositions that arise in utility analysis. This assumption isolates the effect of goals on rational decisions and puts aside the effect of decisions on goals. Finally, agents' goals are assumed to be comparable so that a single scale for utilities is possible. There may be cases in which agents have incomparable goals, but my idealization puts aside such cases.

According to my large idealization about decision situations, there are no conflicts between the various subgoals for rational decisions, and there are no obstacles to meeting those subgoals and the overall goal except, perhaps, uncertainty about options' outcomes. Also, options are certain to be realized if chosen, maximization of utility is feasible, and utility comparisons of options are independent of information provided by options. These idealizations about goals and options are for the decision rule to maximize utility, not for my principles of utility analysis. Their content depends on the type of decision to which they apply. In the context of group decisions, for example, they imply that the costs of communication do not prevent the realization of decision goals. My idealizations put aside the special explanatory factors that arise in certain problem cases and allow me to treat maximizing utility as a goal of rationality.

Under the idealizations, my utility and decision principles advance necessary conditions for rationality. Agents must comply to be rational. I do not claim that compliance with my principles suffices for rationality. Rationality may impose constraints beyond those I formulate. Also, I formulate rationality's requirements using principles that express the motivation of rational agents under the idealizations. Rationality requires having the right motivation for utility assignments and decisions. My principles indicate reasons, not just acceptable results, for utility assignments and decisions. They express goals of rationality.

A principle for ideal cases may express a goal of rationality, although rational agents often have good excuses for falling short. Also, the goal expressed may retain its motivational role in nonideal cases. My principles of rationality express goals of rationality for ideal and nonideal cases alike. They express goals for all cases to which they apply. The absence of obstacles to the goals' attainment in ideal cases makes their attainment mandatory in these cases. In nonideal cases attaining the goals is not mandatory, but the goals still guide decision. In this way my principles for ideal cases have explanatory power in nonideal cases. They are relevant because they express goals of rationality that carry over to nonideal cases. The goals direct the formulation of principles for those cases. By creating an environment that reveals fundamental goals of rationality, my idealizations establish a theoretical framework extendible to nonideal cases.

In nonideal cases agents may have excuses for failing to attain the goals of rationality. Violations of principles expressing them are not necessarily irrational. Although agents often have excuses for non-compliance, rational agents still aspire to comply. They need not put compliance at the top of their priorities; the goals my principles express are not overriding goals. Other goals may compete and take precedence. Nonetheless, rational agents make reasonable efforts, in light of other goals, to pursue the goals of rationality my principles express. Rational agents pursue the goals in ways appropriate to their circumstances and cultivate intellectual habits that help them better pursue those goals.

Kaplan (1996: 36–8) takes a similar view of certain rules of probability. He says that they express a regulative ideal. Although violations may be excused, they are nonetheless mistakes. A rational agent recognizes that he is open to criticism if he violates the rules. Similarly, Levi (1997: 41) says that certain principles of belief express commitments that rational agents have. Although the agents may have excuses for failing to comply, those commitments demand reasonable efforts at compliance. Regulative ideals and cognitive commitments set standards similar to the goals of rationality expressed by my utility and decision principles.[16]

---

[16] Vineberg (1997: 199) interprets the probability law of reflection as an epistemic ideal and uses this interpretation of it to parry objections.

Although idealizations in decision theory are acceptable, one ultimately wants a decision theory without idealizations. A way to make progress toward such a theory is to advance an idealized theory and then remove idealizations one at a time. For example, dispensing with the idealization to quantitative cases is desirable. My theory is more useful if generalized for nonquantitative cases – that is, if the goals expressed by the theory are generalized for cases where the quantities the theory assumes are not available. I would also like to generalize for cases where decision makers are uncertain of their desires, where they fail to consider all the options open to them, and where they have made errors for which decision principles must compensate.[17] In future work I plan to remove at least some of my idealizations. There is already, available to draw on, an instructive body of literature on removing idealizations about the availability of probabilities and utilities, the stability of options' expected utility rankings, and the rationality of agents – for instance, taking these topics in order, Levi (1980, chap. 9), Jeffrey (1983, sec. 1.7), and Sobel (1976).

1.4. EMPIRICISM IN DECISION THEORY

Besides its idealizations, another important feature of my decision theory is its reliance on theoretical entities, that is, entities not directly observable. For example, my decision theory presumes probabilities and utilities, which are not directly observable. How should we introduce such theoretical entities? The prevalent empiricist school of

---

[17] For the last class of cases, it is important to distinguish between a decision in the sense of an option selected and a decision in the sense of a selection process. My standards for the selection process are noncomprehensive. They take present circumstances for granted. My standards for the option selected are comprehensive. They consider whether present circumstances involve mistakes for which some compensation is in order. For instance, a decision maker may have some irrational beliefs about the options he is considering. The standards for the selection process are nonetheless applied with respect to his actual beliefs, but the standards for the option selected consider the beliefs he ought to have as well as his actual beliefs. When comprehensive standards are applied, I say they evaluate the *decision*, not just the *choice* from among options. A choice among options may be good, although the resultant decision is bad because the genesis of the choice has bad elements. Comprehensive standards for decisions look to the origin of the choice as well as the choice itself. My decision theory advances comprehensive standards for decisions rather than noncomprehensive standards for choices. However, its idealizations, which put aside the issue of compensation for errors, make idle the distinction between comprehensive and noncomprehensive standards.

thought divides on this issue. Strict empiricism, or operationism, insists that all concepts be defined in terms of observables. Tolerant empiricism permits in addition concepts introduced by their roles in theories, provided the theories are such that observational evidence for or against them is possible. Philosophy and scientific practice support tolerant empiricism. I adopt its theory of meaning, which I call *contextualism*, because it allows contexts of use to introduce concepts. Contextualism makes available a richer assortment of concepts and consequently yields a richer decision theory than operationism does. In particular, it allows the introduction of probabilities and utilities that explain preferences, whereas operationism is confined to probabilities and utilities that merely represent preferences. No compelling argument about meaning requires decision theory to shackle itself with operationism. So theoretical fruitfulness recommends its rival, contextualism.[18]

This section defends contextualism because some writers, such as Binmore (1994: 181), favor operationism in decision theory. However, because my main topic is rational decision and not decision theory's method, the defense is brief. I do not fully articulate contextualism and the case for it, and I argue for the superiority of its account of concept introduction with respect to only one alternative account, namely, operationism's. I simplify issues and omit many relevant considerations and arguments. My aim is only to outline the case for contextualism.[19]

1.4.1. Empiricist Theories of Meaning

Empiricism, which has widespread support, is best known for epistemological doctrines emphasizing reliance on observation rather than reliance on innate ideas, in contrast with its rival, rationalism. But empiricism also advances a theory of meaning. It sanctions only

---

[18] I assume that some version of an empiricist theory of meaning is correct. Some theorists dispute my assumption. Fodor (1998: chap. 4) argues that empiricism takes concepts to be recognitional and that recognitional concepts violate the principle of compositionality. Because he ascribes to the principle of compositionality, he denies recognitional concepts and hence empiricist theories of meaning. However, there are ways for concepts to be connected with observation without being recognitional in Fodor's sense. Moreover, as Horwich (1998: 35, 154) argues, compositionality by itself imposes no significant constraints on the meanings of words.

[19] I presented this section in Bled, Slovenia, on June 6, 2000, at the year 2000 Bled Conference on Philosophical Analysis and thank the participants for stimulating comments.

concepts introduced via observation.[20] The method of introduction it sanctions is not completely clear. How the empiricist theory of meaning is articulated has momentous consequences for decision theory's scope and power. Let me compare more closely the two previously mentioned empiricist standards for concept introduction – operationism's and contextualism's.

Empiricist theories of meaning may be divided into two kinds. The strict kind holds that all concepts, even theoretical ones, should be defined in terms of the observable – more precisely, what is directly and intersubjectively observable in principle, if we assume no change in the laws of nature and our powers of observation.[21] I call strict empiricism *operationism* after the strict empiricist school by that name.[22] Logical empiricism or positivism and verificationism are also types of strict empiricism.[23]

The nonstrict kind of empiricism allows theoretical concepts to be introduced in ways that tie them more loosely to the observable. It allows theoretical terms to be introduced, not necessarily defined, by claims connecting them with observational and previously introduced nonobservational terms. Nonstrict empiricism accepts theoretical terms that are primitive. These primitive theoretical terms acquire meaning not through definition but through use in a context. Instead of being defined by necessary and sufficient conditions of application, a term is used in a sentence or set of sentences. Its usage allows an uninitiated person to grasp the concept it expresses. Such contextual introductions are common for theoretical concepts. The concept

---

[20] Internalist versions of empiricism may say the path to meaning is sensation rather than observation. My brief treatment of empiricist tenets about meaning neglects some varieties of empiricism.

[21] By definition, I mean definition in the philosophical sense. Accordingly, a word's definition provides a synonym that explains the word's meaning. For example, a vixen by definition is a female fox. See T. Yagisawa's entry for "definition" in Audi (1995).

[22] For a recent account of operationism, see the entry by G. Hardcastle in Garrett and Barbanell (1997).

[23] Operationism condones definitions using observational terms, logical terms, and other unspecified auxiliary terms, perhaps the connective for counterfactual conditionals. Logical empiricism is more precise and more demanding than operationism. It requires definitions using only observational and logical terms – the constants of standard extensional logic – and so excludes, for example, the connective for counterfactual conditionals. Verificationism claims that an expression's meaning is a method of verifying its application. An operational definition of an expression may formulate such a method.

of an electron, for instance, is introduced via the theory of the atom. I call the theory of meaning that supports these introductions *contextualism*.[24]

What are the consequences for decision theory of adopting the strict empiricist theory of meaning? If operationism is adopted, decision theory must forgo group and intrinsic utility analyses. Group utility analysis uses interpersonal utility, which lacks a satisfactory operational definition because its connection with group behavior is too indirect. Intrinsic utility analysis uses intrinsic desires and aversions, which also lack satisfactory operational definitions because their connection with behavior is too indirect. Group and intrinsic utility do not pass muster in the operationist's camp, and the operationist dismisses them as meaningless. Furthermore, under operationism, expected utility analysis loses its power to explain behavior. Behavior is used to operationally define subjective probability and utility. Hence they cannot explain behavior via expected utility analysis. Therefore, operationism dismantles multidimensional utility analysis and, in particular, blocks my approach to decisions made for others.

Because of the consequences, one should not adopt operationism unless the evidence for it is compelling. However, a host of writers has discredited the operationist standard of meaning because it is too stringent.[25] Contextualism offers a much more plausible standard of meaning. Although a full exposition of contextualism is beyond this book's scope, its credentials are well established in the literature on theories of meaning.

---

[24] Lewis (1998: chap. 9) does not endorse an empiricist criterion of meaning but attempts to explicate the class of empirical statements in a way that includes scientific theories as well as observation statements. He takes empirical statements as statements partly about observation. Being partly about observation is vague and leads to various empiricist criteria depending on how the vagueness is resolved. One disambiguation says, "A statement S is partly about observation, in the sense of partial supervenience, iff observation is evidentially relevant to S" (p. 151). This disambiguation, also advanced by Skyrms (1984: 14–19), yields an explication of the empirical in the spirit of contextualism, but with some differences. Contextualism advances a standard for the introduction of a meaningful term of any sort, whereas Lewis's explication of the empirical advances a standard for statements only, and for the statements themselves rather than for their introductions.

[25] Hempel (1965: chaps. 4, 5) is a trenchant critic of the strict empiricist, or operationist, standard of meaning. He (1966: sec. 3.5) adheres to a nonstrict version of empiricism that demands testability rather than definition in terms of observables. He takes a theoretical term as meaningful if some sentences containing it are testable. Here by "testable" he does not mean subject to conclusive confirmation

Ramsey (1931, sec. IX.A) proposes a contextualist theory of meaning. He holds a "theory theory of meaning" according to which a theory in which a term is used grounds its meaning. Davidson's theory of meaning (1984, chap. 2, esp. p. 23) is another version of contextualism. According to it, the meanings of words are illuminated by the truth conditions of sentences containing them. My version of contextualism does not insist that meanings be strictly derivable from theories or truth conditions; rather, their extraction may invoke human psychology, not just logic, and may rely on nonverbal features of context such as demonstrations.[26]

Horwich (1998) advances a theory of meaning according to which the meaning of a word derives from its use. While presenting his theory, he considers the distinction between a word's explicit definition (pp. 79–80), which states a synonym for the word, and a word's implicit definition (pp. 131–42), which simply makes assertions containing the word. His theory of meaning easily grounds implicit definition, which is contextualistic. Assertions containing a word may indicate a usage for the word and so make it meaningful. Horwich's use theory of meaning also supports contextualism.

A criterion for concept introduction is just part of a complete theory of meaning. A complete theory elucidates the nature of concepts, their relationship to linguistic expressions, and much more. Contextualism addresses concept introduction only. It is neutral about other components of a complete theory of meaning. Most controversies dividing theories of meaning concern these other components. Contextualism

or disconfirmation. A sentence is testable in the relevant sense roughly if it is possible to acquire intersubjective observational evidence that raises or lowers its probability.

Using testability as a criterion of meaning is compatible with contextualism. Only some forms of contextualism reject the testability requirement, and some of them reject only the intersubjective version of the requirement. In any case, the requirement is weak and excludes very little of interest for decision theory. The decision theory advanced here uses only terms that meet the testability requirement. In particular, probability and utility taken as primitives meet the testability requirement, given their connection with overt behavior.

[26] Because of the role of human psychology, a contextualist introduction of a term may succeed even if the introductory remarks are not all true. Also, contextualist methods of introducing a term may draw on human psychology in different ways according to the term's type. Kripke (1980) and Putnam (1975) describe contextualistic methods of introducing names, including names of natural kinds, via reference-fixing acts of various sorts. These methods differ from introductions of technical terms via a theory.

makes a simple point about the distinction between meaningful and meaningless expressions, namely, that some expressions lack operational definitions but are meaningful in virtue of introductory remarks using those expressions. In adopting contextualism, I assume, in particular, the efficacy of introductions of theoretical terms that work by assertions involving the terms but fall short of operationist definitions. I do not investigate the type of meaning these theoretical terms acquire and the mechanisms by which their introductions invest them with meaning. That is a job for a complete theory of meaning.

In presenting decision rules and principles of utility analysis, I introduce probability, utility, and other technical concepts as primitive concepts. I do not define them in terms of observables but introduce these concepts via the theories of which they are part. For example, the concepts of probability and utility are introduced via the theories of probability, utility, and decision – as in Schick (1991: 39–40). I rely on the reader's ability to acquire the technical concepts given their introductions. Although the introductions are not as precise as definitions, they make the technical concepts serviceably clear.

Introducing theoretical concepts as primitives is not a lazy theoretician's tactic. A contextualistic introduction of a theoretical concept often requires more work and finesse than an operational definition of a rough-and-ready substitute. Achieving sufficient clarity without operational definition generally demands ingenuity and insight. Contextualist introductions of probability and utility draw on the subjects' rich history, which is filled with hard-won distinctions, innovative principles, and clever theorems.

Probability and utility taken as primitives meet all plausible criteria of meaning. They are connected via theory with observables such as behavior even if they are not defined in terms of observables. The theories of probability and utility and the mechanisms of human psychology fill out the concepts. The same contextualist account of meaning that supports theoretical concepts of physics also supports theoretical concepts of decision theory. We need not be stricter about concepts in decision theory than in physics.

1.4.2. Contextualism's Advantages

Although operationism is a mistaken theory of meaning, some theorists think that science should nonetheless restrict itself to operationally defined terms. They argue that even if other terms are

31

meaningful, lack of operational definition prevents their incorporation in science. Science has testability requirements that can be met only if all its vocabulary is operational. Decision theorists in the empirical sciences often adopt the operationist restriction for such methodological reasons. Decision theorists in philosophy often adopt it because of the methodological proclivities of the empirical sciences with which their field is allied. As a result, decision theory generally eschews theoretical concepts not defined in terms of observables.

Does decision theory benefit methodologically from the operationist restriction? To simplify an evaluation, I concede that operationism generates clarity.[27] I argue that, despite that advantage, contextualism is superior. Although operational definition achieves clarity, decision theory suffers on balance from operationism. As this section argues, operationism discards many advantages besides multidimensional utility analysis. In contrast, contextualist decision theory secures those advantages by discarding the operationist restriction. Contextualism forgoes the clarity of operational definition primarily to facilitate theory construction. Condoning concepts not defined in terms of observables allows the formation of theories better than their operationistic counterparts. Contextualism appreciates clarity but is willing to sacrifice observational clarity for theoretical advances.

Because contextualism allows a richer stock of theoretical concepts than operationism, contextualist decision theory is significantly richer than operationist decision theory. There are three reasons to opt for this enrichment that together outweigh the loss of clarity that comes from forgoing definitions in terms of observables. First, the enriched theory offers deeper, more systematic justifications of decisions – that is, explanations of their rationality. Its additional conceptual resources permit going behind phenomena for explanations and generalizing and unifying explanatory principles. Second, the enriched theory is more accurate because to deal with difficult cases precisely it is necessary to appeal to entities not recognized by operationism. Third, the enriched theory is more comprehensive and powerful. It uses its richer set of primitive concepts to formulate new principles of utility analysis and, by means of these new principles, resolves recalcitrant decision problems, such as the problems that

---

[27] Some observational concepts, such as the concept of the color orange, are vague and so may be unclear.

arise when members of the professions make decisions for their clients.[28]

The advantage of contextualist decision theory I emphasize most is its explanatory power. For example, the principle to maximize utility has normative force only if utility is not defined as a quantity maximized by rational choices. If utility is defined in terms of an agent's rational choices, as in revealed preference theory, then utility maximization cannot explain the rationality of the agent's choices.

1.4.3. Operationist and Contextualist Explanations

Is there any reason not to take probabilities and utilities as primitives in decision theory, or to resist using them to explain rational preferences? In general, is there any reason for the sciences and other disciplines to adhere to operationism's strict standard of meaning despite operationism's failure as a theory of meaning? No, the operationist standard of meaning is too demanding. It dismisses virtually all the theoretical concepts of science. Moreover, there is no methodological reason to restrict the truth-seeking disciplines to operationally defined concepts despite the severity of the operationist theory of meaning. Rejecting the theoretical concepts of science thwarts explanations that go behind the phenomena. Good scientific explanations need theoretical concepts not available under operationism. Decision theory similarly needs theoretical concepts not available under operationism and should be weaned from its operationist origins.

Operationist science is aware that, by insisting on definition in terms of observables, it dispenses with entities useful in explaining phenomena. As a final effort to defend a restriction to operationally defined terms, it plays down the significance of explanations. Typically, it criticizes our common conception of explanation and claims that we are mistaken to press for explanations in the ordinary sense. It holds

---

[28] Binmore (1994: 50–1, 105, 169, 180–1) acknowledges that economics, defining utility in terms of choices, forgoes explanations of rational choices. Although he recognizes the value of explanations, he is reluctant to abandon operationist definitions to attain them. He leaves explanations of rational decisions for a yet to be constructed "fully articulated theory of rationality." My project is to make progress toward such a theory. I abandon operationist definitions of utility in terms of choices but retain economics' methods of inferring utility from choices. I make utility psychological without identifying it with pleasure or satisfaction. I let it encompass the variety of factors that motivate choice without reducing it to choices.

that such explanations involve unwarranted anthropomorphizing; they attempt to make phenomena understandable *to us*. Insofar as explanation is warranted, there is nothing more to it than deducing the phenomenon to be explained from general laws and particular facts.[29] The concept of causality that is often thought to be part of the concept of empirical explanation is dismissed as a will-o'-the-wisp since it cannot be defined in terms of observables.

The problems with this operationist account of explanation are well known. It does not adequately distinguish explanatory factors from symptoms, as in the famous case of the flagpole and its shadow. Given the angle the sun makes with the earth's surface, the length of the flagpole may be deduced from the length of its shadow (and vice versa). But the length of the shadow does not explain the height of the flagpole. It is the other way around. Also, the operationist account of explanation relegates explanations, like the famous one concerning paresis and latent syphilis, to the category of incomplete explanations. Because latent syphilis is a necessary but not a sufficient condition for paresis, latent syphilis is not counted an explanation of paresis. The operationist account of explanation fails to recognize that an explanation may be a complete account of one causal factor involved in the production of a phenomenon even if it does not provide the complete cause of the phenomenon. Operationism does not recognize the varieties of explanatory success. In particular, it misrepresents the type of explanation furnished by scientific theories that use idealizations. These theories furnish complete accounts of some causal factors behind the production of certain phenomena although they do not treat all the relevant causal factors. They have a type of completeness that the operationist account of explanation does not recognize.[30]

Operationist science takes these drawbacks in stride, but contextualist science does not. It is more demanding than operationism in the area of explanation. Not satisfied with deducibility from law, its goal is causal explanation.[31] Also, it seeks systematic, deepgoing explanations.

---

[29] This is a brief description of Hempel's (1965: pt. IV) covering law model of explanation. I bypass details of Hempel's elaborate treatment of explanation because the simple points made here do not require their examination.

[30] See Scriven (1959). Hempel's account of explanation faces such objections because it is adamantly noncausal. Although his account of meaning is not strictly empiricist, he harbors strict empiricism's doubts about causality.

[31] A full causal explanation of an indeterministic phenomenon reveals the objective probabilities or propensities that govern it.

It wants to explain diverse phenomena by means of a few simple, inter-active, deep-level laws, as, for example, Newton's theories of motion and gravity explain laws governing the movements of falling objects, the planets, the tides. These systematic explanations can be obtained only by going behind the phenomena to fundamental laws concerning theoretical entities. System appears only at the theoretical level.

Contextualist decision theory seeks deepgoing systematic causal explanations of rational preferences (although it recognizes noncausal types of explanation too). In particular, to obtain this kind of explana-tion, it seeks general principles of utility analysis and seeks to avoid restrictions or boundary conditions. Generality adds depth to explana-tions. Out of regard for generality, my contextualist decision theory formulates principles of utility analysis so that they apply not only to the options of a decision maker, but to all events and to events given conditions, and to various types of utility.

Although contextualist science seeks to avoid restrictions, it is willing to adopt idealizations. An idealization serves to isolate causal factors, whereas a restriction merely eliminates troublesome cases, as Section 1.3.1 explained. Idealizations play a crucial role in partial explanations. Because contextualist science has demanding standards of explanation, the explanations it can provide are often only partial and involve idealizations. Even with the help of a richer set of theo-retical entities, it may not be able to produce a full causal explanation of a phenomenon. It may have to settle for an explanation of the effect of some causal factors contributing to the phenomenon, relying on idealizations to isolate those factors and to describe their role. For example, in explaining why a cannonball traveled so many meters, a partial causal explanation may cite its velocity as it left the cannon's muzzle, gravity, and the cannon's angle with the earth's surface. Other relevant causal factors such as air resistance may be omitted. They may be idealized away to enable the explanation to give a precise account of the causal role of velocity, angle, and gravity.

I do not offer an account of partial explanations that isolate causal factors using idealizations. An account in terms of a probability increase for the event to be explained is promising, but the issues are complex. Fortunately, for my purposes, there is no pressing need for an account of partial explanations. Partial explanations are familiar and understood even without an account of them.

Contextualist decision theory must also often temporarily settle for partial explanations. Idealizations in decision theory are needed to isolate factors involved in the explanation of rational preferences. Rationality is a complex matter. It takes into account, for instance, limitations on an agent's ability to compute expected utilities. When expected utility analysis is used to explain rational preferences, such limitations and other factors are put aside by means of idealizations. Expected utility analysis, therefore, explains rational preferences, but only partially.

Operationist theories are inimical to many common idealizations. The operationist standard of meaning dismisses ideal entities, such as frictionless bodies and point masses. Because these ideal entities defy the laws of nature, they are not definable in terms of observables. There is nothing observable with which to define them, because their existence assumes new laws of nature, and observability in the relevant sense preserves the laws of nature. The rejection of ideal entities thwarts partial explanations that use such ideal entities, however. So operationism undercuts partial explanations.

Decision theory, in particular, would be severely impoverished if ideal entities were discarded. For example, it is common and useful to assume the existence of quantitative probabilities and utilities. But these psychological quantities may not actually exist. They may be only ideal entities. In this case a thoroughgoing operationist must dispense with them – making accuracy, generalization, and systematic explanation much harder to achieve. Ideal entities, although fictions, facilitate partial explanations of phenomena.

The foregoing considerations concerning explanation and theory formation provide a strong general case against adopting an operationist restriction for methodological reasons. In philosophical decision theory, where the treatment of decisions is normative rather than descriptive, the case against a restriction to operational terms is even stronger. Philosophical decision theory is unencumbered by the special demands of the empirical sciences, such as economics, that use normative decision theory to help describe, predict, and explain the behavior of people (although people are imperfectly rational). Only a carefully regulated normative theory helps the sciences meet their goals. Because of the sciences' methodological constraints, the normative theory must be applicable in ways that are intersubjectively verifiable by observation. Both philosophy and the empirical sciences aim at the truth, but being different disciplines they have different methodological objectives. Philosophical decision theory does not have the methodological constraints of empirical decision theory. It has the freedom to formulate rules of rational decision making without any objective besides truth. It does not have the empirical sciences' additional objective of testability by intersubjective observation. It may use theoretical entities, such as interpersonal utilities, that are not measurable accurately enough to meet the empirical sciences' testability requirements.[32]

Consider ethics, another normative discipline. Rejecting nonoperational terms lays a heavy hand on ethical theory. It rules out utilitarianism on methodological grounds because the utility principle invokes interpersonal utilities. Plainly, it is inappropriate to insist on operational definitions of ethical concepts. For instance, it is inappropriate to insist on a definition of moral obligation in terms of observables. Attempted definitions commit the naturalistic fallacy. It is similarly inappropriate to insist on operational definitions of philosophical decision theory's normative concepts. Attempted operational definitions of rationality also commit the naturalistic fallacy. They attempt to define the normative in terms of the nonnormative.

Insistence on operational definitions is also inappropriate for the nonnormative concepts of philosophical decision theory. Take degrees of belief and desire. A common argument for defining them in terms of observables cites the sciences' need for intersubjective verifiability by observation. This argument has no force in philosophical decision theory because the discipline is not subject to the sciences' testability requirements. A philosophical decision theory need not be intersubjectively applicable. At most, it must be applicable by the decision maker. Consequently, it would be enough, for individuals making decisions for themselves, if degrees of belief and desire were defined in terms of the introspectible. They would not have to be defined in terms of the intersubjectively observable.

Notice that I reject only the imposition of an operationist standard of *meaning*, a standard calling for *definition* of all concepts in terms of observables. Operationism is a false theory of meaning, and methodological reasons in decision theory and science are insufficient for nonetheless requiring all concepts to be operationally defined. I am not rejecting the sciences' testability requirements and their consequences

---

[32] My points about economics' methodology apply to that discipline only as traditionally conceived. Some economists – for example, Rubinstein (1998: sec. 11.2) – justify their theories not in terms of empirical goals but in terms of conceptual goals. They even compare their methods to philosophy's.

for scientific methods. In the sciences, for the sake of intersubjective testability, it may be appropriate to require that all entities, including theoretical entities, be assigned observational methods of detection or measurement. In other words, it may be appropriate to require that they be operationalized in an inferential sense. But even if the special concerns of the empirical sciences motivate a demand for methods of inferring theoretical entities from observables, they do not motivate a demand for definitions of theoretical entities in terms of observables. They, at most, motivate a requirement of operationist inference and not operationist meaning. Furthermore, the motivation for that requirement does not extend to philosophical decision theory because this discipline does not have the same testability requirements as the sciences.[33]

The decision theory I advance disowns operationistic objectives. It is less demanding than operationism about the introduction of theoretical concepts. It is contextualistic. Moreover, being philosophical and unbound by the methodological constraints of the empirical sciences, it need not provide operationistic substitutes for its theoretical terms for the sake of disciplinary testability requirements. Decision theory gains enormously by adopting contextualism rather than operationism. The richer set of concepts contextualism provides allows the theory to blossom, as the following chapters show.[34]

---

[33] An operationist requirement of measurability by intersubjective observation might be motivated in the sciences by the goal of making all scientific theories intersubjectively testable. Here the motivation is related to standards of evidence rather than standards of meaning, and the requirement puts aside certain types of evidence rather than certain types of concept. It puts aside evidence that is too imprecise to count as measurement and too reliant on introspection to count as intersubjective. The requirement is controversial for the sciences and clearly does not apply to philosophical decision theory.

Goldman (1996, 1997) argues against imposing on the sciences an intersubjective testability requirement, one demanding that scientific evidence be intersubjective. He claims that such a requirement lacks epistemic justification. It ignores the evidence of introspection. Putting aside the evidence of introspection, he remarks, may even undermine the foundation of commonly accepted types of scientific evidence concerning the mental states of people. I leave the issue open. Perhaps there are pragmatic reasons for intersubjective testability in the sciences, or perhaps convention makes it a hallmark of science. I defer to the scientists who insist on intersubjective testability. However, the intersubjective testability requirement does not govern philosophical decision theory. Philosophical decision theory has no epistemic, methodological, or conventional reason to ignore any type of evidence, including the evidence of introspection.

[34] Some authors have noted differences in the approaches to decision theory of French and American theorists. For instance, the French economist, Allais, pub-

This book elaborates a contextualist theory of decision richer, more systematic, and more accurate than its operationist rivals. For unity, all forms of utility analysis are placed within my multidimensional framework. The various forms of utility analysis are all justified as ways of assessing options with respect to basic intrinsic desires and aversions. Chapters 2, 3, 4, and 5 present intrinsic utility analysis and expected utility analysis, and their two-dimensional combination. Chapter 6 presents a new, independent form of analysis, group utility analysis, and

lished an article in 1953 entitled "Foundations of a Positive Theory of Choice Involving Risk and a Criticism of the Postulates and Axioms of the American School" and the American psychologist, Lopes, published a paper in 1988 entitled "Economics as Psychology: A Cognitive Assay of the French and American Schools of Risk Theory."

The American school of decision theory was founded by the mathematician von Neumann and the economist Morgenstern in *Theory of Games and Economic Behavior* (1944), and by the statistician Savage in *The Foundations of Statistics* (1954). One of the main features of this school is its definition of probabilities and utilities in terms of a preference ranking of options involving chance. Another main feature, added by Pratt (1964) and Arrow (1965, 1970), is its definition of degrees of aversion to risk in terms of a preference ranking of options involving chance.

The French school of decision theory, founded by Allais in the article mentioned and recently advanced in Allais and Hagen (1993), is less clearly characterized. Sen (1985) points to its rejection of the definition of utilities in terms of preferences between options, and its use of utilities to explain such preferences. On the other hand, Lopes (1988) points to its rejection of the definition of degrees of aversion to risk in terms of preferences between options. She notes that the French school treats risk as a consequence of an option involving chance and treats a decision maker's degree of aversion to the risk involved in an option as a factor determining the utility for her of the overall outcome of the option.

I think that the main difference between the French and American schools lies deeper than the features that Sen and Lopes mention. It is a philosophical difference concerning theoretical quantities such as utilities and degrees of aversion to risk. The American school insists on defining these theoretical quantities in terms of readily observable preferences. But the French school allows introductions of the quantities that make their connection with readily observable preferences less direct. The American school derives its position on theoretical quantities from the operationism that flourished in the United States during the first half of this century. It stresses the clarity achieved by defining theoretical quantities in terms of observables. The French school, on the other hand, derives its more liberal position from traditional scientific methodology. It stresses the explanatory fruitfulness of theoretical quantities that lie behind observables and are not definable in terms of observables.

Perhaps the American and French schools can be reconciled by showing that they undertake different research programs and that the position of each school on theoretical quantities is appropriate for the research program it undertakes. The

articulates the three-dimensional analysis achieved by combining it with intrinsic utility analysis and expected utility analysis. Chapter 7 demonstrates the power of multidimensional utility analysis by applying it to decisions made for others. Many of these complex decisions require the conjoint application of intrinsic, expected, and group utility analyses. Chapter 8, the final chapter, summarizes results and briefly entertains topics for future research. An appendix verifies the consistency of the various methods of utility analysis. It shows that they compose a unified, multidimensional type of utility analysis, the foundation of which is intrinsic utility analysis.

American school emphasizes the mathematical representation of preferences, and for this research program an operationist restriction on theoretical quantities is appropriate. The French school, in contrast, emphasizes the explanation and justification of preferences, and for this research program a contextualistic liberality concerning theoretical quantities is appropriate. This book's research program is the construction of a decision theory with principles that justify preferences. In this respect it is more closely allied with the French than the American school.

2

Intrinsic Utility Analysis

One traditional form of utility analysis assesses an option's utility by tallying the utility derived from the option's realization or frustration of the agent's goals. For example, reading *War and Peace* may achieve the goal of absorbing Tolstoy's great novel but frustrate the goal of painting the house. Each goal's weight depends on its realization's utility. The utility of reading the classic is then a sum of weights positive for goals it realizes and negative for goals it frustrates. This chapter refines and elaborates the method of utility analysis, which I call *intrinsic utility analysis*.

Section 1.1 applied intrinsic utility analysis to the social utility of a new safety standard. The analysis used some of society's goals: health, security, and prosperity. Such goals form a dimension of analysis for utilities. Intrinsic utility analysis relies on the dimension of goals – that is, considerations appealing directly to goals. Other forms of analysis rely on other dimensions, or types of consideration. This chapter explains how intrinsic utility analysis uses goals to calculate utilities. Later chapters introduce additional methods of utility analysis.

Intrinsic utility analysis is named after its elements. It reduces utilities to *intrinsic utilities*. Intrinsic utility is a new, noncomprehensive type of utility, so named because it assesses only the intrinsic features of objects of desire and aversion. Section 2.2.1 introduces intrinsic utility as part of an explanation of intrinsic utility analysis.

Intrinsic utility analysis can decompose intrinsic utilities and also utilities of other types. This versatility is critical for my purposes. The decision principle to maximize utility, my touchstone, uses comprehensive utility, or utility all things considered. I rely on intrinsic utility analysis's ability to calculate comprehensive utilities from intrinsic utilities. Section 2.3 presents a formula for the comprehensive utility of a possible world. It uses intrinsic utilities of objects of basic intrinsic attitudes, which are causally fundamental intrinsic desires and aversions.

According to the formula, *a world's comprehensive utility is the sum of the intrinsic utilities of objects of basic intrinsic attitudes it realizes.* Comprehensive utilities of worlds suffice for decisions given certainty about options' outcomes. However, complications concerning uncertainty, addressed in Chapters 3 and 4, postpone intrinsic utility analysis's general application to decisions. I wait until Chapter 5 to explain intrinsic utility analysis's extension to options' comprehensive utilities.

2.1. INTRINSIC DESIRE AND AVERSION

Intrinsic utility involves a noncomprehensive type of assessment attending to intrinsic qualities only. This section explains the type of assessment and formulates principles that govern it. The next section uses this type of assessment to introduce intrinsic utility.

Because the chapter's objective is assessment of possible worlds, I attend especially to the requirements for it. Intrinsic utility assesses a possible world with respect to an agent's goals. To ensure conformity with the principle of pros and cons, it uses goals that are basic. It separates considerations for or against a world according to the basic intrinsic desires and aversions that the world realizes – that is, whose objects would be true if the world were realized. Realizing a basic intrinsic desire is a pro, and realizing a basic intrinsic aversion is a con. Adding the utilities of these pros and cons yields the world's intrinsic utility. The basic intrinsic desires and aversions are defined so that they divide considerations without omitting or double-counting anything relevant. Intrinsic attitudes are the topic of Section 2.1.1, and basic intrinsic attitudes the topic of Section 2.1.2.

2.1.1. Intrinsic Attitudes

An agent may assess an option according to the extent to which it would advance him toward his goals. His goals provide a convenient way of separating pros and cons for an option. As I understand a goal of an agent, it is an object of an *intrinsic* desire, a desire for something exclusively because of its intrinsic qualities, and not also because of its nonintrinsic, or extrinsic, qualities. For example, an agent may desire to run because of running's intrinsic qualities, that is, its inherent and non-relational qualities. His desire may be independent of running's health benefits, for instance. Then his desire to run is intrinsic.

An agent's goal is also an end, something that the agent desires for its own sake, not merely as a means to other things. Pleasure is typically an object of intrinsic desire and an end, whereas wealth is typically an object of nonintrinsic, or extrinsic, desire and desired as a means. I do not say that a person who desires something has just one desire for it that is either intrinsic or extrinsic depending on whether the desire arises exclusively from the thing's intrinsic features, and that is either final or not depending on whether the thing is desired as an end. Some things, such as health, are objects of both intrinsic and extrinsic desire and are desired both as ends and as means to other things. An intrinsic desire for health arises exclusively because of health's intrinsic qualities, and an extrinsic desire for health arises in part because of the extrinsic features of health. Furthermore, a person typically desires health not only for its own sake but also as a means to other things, such as productivity.[1]

To make the foregoing more precise, I examine more closely objects of desire and their assessment. On my view, both intrinsic and extrinsic desire attach to propositions. I take propositions to be the basic bearers of truth values, the meanings of declarative sentences. They have structures incorporating concepts. Their structures are similar to the syntactic structures of sentences. They may be negated, conjoined,

---

[1] An intrinsic desire is not a desire that is intrinsic but rather a desire whose grounds are intrinsic features of its object. The term *intrinsic desire* comes from Brandt (1979: 111). He uses it for a desire assessing intrinsic qualities in part. I use it for a desire assessing intrinsic qualities exclusively.

Feldman (1986: sec. 2.2.5) treats intrinsic utility. He is interested in intrinsic utility as a measure of moral value, whereas I am interested in it as a measure of intrinsic desire. Still, the two types of intrinsic utility are structurally similar. His principles of intrinsic utility, applied to rational degree of intrinsic desire, are compatible with mine. Quinn (1974) considers problems with the extension of intrinsic value from worlds to arbitrary propositions. Analogous problems arise for the extension of intrinsic utility from worlds to arbitrary propositions. This chapter targets calculation of the intrinsic utilities of worlds only.

Korsgaard (1983) distinguishes intrinsic value and final value: value from inherent qualities and value as an end. I distinguish intrinsic desire from final desire, or desire as an end, along similar lines. Every final desire is an intrinsic desire, and also a basic intrinsic desire. But not every intrinsic desire is basic, and so a final desire. An agent may lack a basic intrinsic desire, intrinsic desire, and even extrinsic desire that q, and yet have a nonbasic intrinsic desire that $p \& q$, because of a basic intrinsic desire that p. In this case there is a sense in which the agent does not desire p & q for its own sake, but for the sake of p. The desire that $p \& q$ is intrinsic but not final. Later this section presents more fully the distinction between basic and nonbasic intrinsic desires.

or disjoined to form compound propositions. However, they are not sentences. Different sentences may express the same proposition.[2]

According to the theory of direct reference, which I adopt, some propositions have concrete individuals as components. The proposition that Socrates lived in Athens, for instance, has Socrates as a component. It is not exclusively composed of concepts. The proper name "Socrates" does not express a concept but refers directly to Socrates.

Evaluations of acts focus on propositions having agents as components. Imagine that Socrates considers fleeing Athens. He assesses the proposition that *he* flee Athens. One factor in his assessment is his desire to honor his obligations to Athens. That desire's object is the proposition that *he* honor his obligations to Athens. His deliberations concern propositions involving him.[3]

Some principles of utility analysis attend to propositions that are maximal and consistent. A maximal proposition entails, for every proposition, that proposition or its negation. A consistent proposition entails no contradiction. In these characterizations, the entailment relation depends on logical relationships in a broad sense, that is, a priori relationships. I call maximal consistent propositions *possible worlds*. These possible worlds are not concrete, although one of them, the actual world, is realized. They yield an account of broadly logical or a priori possibility and necessity. That account regulates utility analysis. I maintain neutrality about the type of possible world serving as the foundation of metaphysical possibility and necessity, and of counter-

---

[2] Frege and Russell hold the view that propositions are structured. Stalnaker (1984) advances a rival view that propositions are sets of possible worlds.

[3] For an introduction to the literature on direct reference, and other theories taking propositions as structured, see Salmon and Soames (1988), Kaplan (1989), and Crimmins (1992). To make objects of attitudes, such as belief and desire, more fine-grained than sets of possible worlds, Lewis (1983: chap. 10) takes them to be properties, that is, sets of individuals, including possible individuals. He compares his approach to taking the objects of attitudes to be sets of centered possible worlds, where a centered possible world is a pair consisting of a world and an individual in that world (sec. X). He also argues that taking propositions as sets of agent-centered worlds leaves Bayesian decision theory intact (sec. XI). I believe he is substantially correct about the effect on decision theory. But I hold that taking an option's outcome as a structured proposition or set of centered worlds requires some restrictions on standard utility laws. The phenomenon of direct reference, and other related phenomena that make the individuation of thoughts external, make it possible for ideal agents to have incomplete knowledge of the propositions they entertain. Utility laws must adjust for their ignorance, as Section 3.2.3 explains.

factual conditionals. Although later discussions of counterfactual conditionals have in mind that type of possible world, my principles of utility analysis depend primarily on the possible worlds behind logical possibility and necessity. So they are my main concern.

Taking propositions as structured entities makes them more fine-grained than taking them as sets of possible worlds. For instance, many structured propositions are logically true, and so true in all possible worlds. The set of all possible worlds does not discriminate among them. This is so whether possible worlds are taken as concrete objects, as in modal realism, or are taken as maximal, consistent (structured) propositions. Under the latter view, logically true propositions are true in all possible worlds; they are entailed by every maximal, consistent proposition. So they are not distinguished by the sets of possible worlds where they are true.

Although intrinsic and extrinsic desires alike attach to propositions, they assess propositions in different ways. An intrinsic desire evaluates a proposition's realization in terms of its logical consequences, in isolation from its causal consequences and its total outcome, whereas an extrinsic desire evaluates a proposition's realization all things considered. The strength of an extrinsic desire that p depends on the causal consequences of p's realization and, in general, the outcome if p were realized. The strength of an intrinsic desire that p depends only on p's logical consequences, and not also on the causal consequences or total outcome of p's realization. Take being wealthy. It does not receive a high rating in virtue of its logical consequences but rather in virtue of its causal consequences and total outcome. It is not valued for its own sake but as a means to other things. It is the object not of a strong intrinsic desire but of a strong extrinsic desire.

In taking the intrinsic features of a proposition's realization as its logical consequences, I mean its logical consequences in a broad sense. These consequences include logical consequences in the formal sense but, in addition, everything else entailed by the proposition. They are all the proposition's a priori consequences. By a priori consequences, I do not mean consequences knowledge of which is temporally or conceptually prior to experience but rather is inferentially independent of experience, that is, independent of experiential premises. Notice that these consequences are epistemically characterized, not metaphysically characterized by necessity, essences, inseparability, or the distinction between qualitative and relational properties. My view of intrinsic desire makes it neutral with respect to most metaphysical issues.

Some theorists say that intrinsic desire puts aside that to which a proposition serves as a means. This needs clarification. When one puts aside the extrinsic features of a proposition's realization to obtain the strength of an intrinsic desire for the intrinsic features of its realization, one puts aside its causal consequences but does not put aside its logical consequences. For instance, to obtain the strength of an intrinsic desire that p & q, one does not put aside q although p & q is a means to q. One puts aside only the nonlogical consequences of p & q. Intrinsic desire puts aside causal consequences, but attends to logical consequences. A proposition's realization is a logical means to its logical consequences, but it is not a causal means to its logical consequences. Intrinsic desire does not put aside that to which a proposition's realization is a logical means, but it does put aside that to which a proposition's realization is a causal means.

Normally, one's intrinsic desires do not vary with one's empirical information, whereas one's extrinsic desires do. An intrinsic desire that p springs from an assessment of the logical consequences of p's realization, which are independent of empirical information. Only in exceptional cases does new information affect an intrinsic desire by casting its object's logical consequences in a new light so that their assessment changes. New information does not require a change in intrinsic desires, but intrinsic desires are revisable, and new information may prompt a revision.

My idealizations concerning mental capacities mitigate the effect of information on intrinsic desires. For ideal agents, new information never casts the objects of intrinsic desires in a new light; new information does not furnish grounds for reassessment. To some it may seem that rationality then requires constancy; however, others may hold that new information may still prompt reevaluation without irrationality. I do not try to settle this issue. Section 1.3.2's idealization that agents have stable goals makes it moot. Given the idealization, one may take an ideal agent's intrinsic desires to be independent of information.

The idealizations also permit postponement of an investigation of the way uncertainty about logical matters affects the formation of intrinsic desires. A person may form an intrinsic desire – for pleasure, say – knowing that its object's unknown logical consequences are irrelevant. In some cases, however, lack of logical perspicuity impedes intrinsic desire. A person's uncertainty about logical matters may, for instance, block the formation of an intrinsic desire that he experiences pleasure if Fermat's last theorem is correct. He may not realize that this intrinsic desire's object is equivalent to his experiencing pleasure. The idealizations table such complications by positing logical omniscience. Although my account of intrinsic desire is general, I apply it to ideal agents only.

The opposite of intrinsic desire is intrinsic aversion. Intrinsic aversion is analogous to intrinsic desire. It goes by logical consequences, too, and is therefore similarly independent of information. Between intrinsic desire and aversion is intrinsic indifference. It has the same scope and type of independence from information. To speak generally of intrinsic desire, aversion, and indifference, I use the term intrinsic attitude.

I introduce a new type of utility, intrinsic utility, that considers all intrinsic attitudes, not just intrinsic desire. It goes against ordinary usage to call the objects of intrinsic aversion and indifference goals. So strictly speaking intrinsic utility analysis breaks down utilities according to objects of intrinsic attitudes, not just goals. I say it analyzes utilities according to goals just for brevity.

Different types of desire and aversion evaluate a proposition more or less broadly. The considerations affecting a proposition's assessment may encompass its logical consequences, its causal consequences, or the entire outcome of its realization. Intrinsic desire narrows the scope of evaluation to a proposition's logical consequences. Propositions desired both intrinsically and extrinsically receive positive evaluations both when the scope of evaluation is narrow and when it is broad. For example, take a desire to be healthy. As an intrinsic desire, it takes account of just the logical consequences of the proposition that one is healthy. As an extrinsic desire, it also takes account of the causal consequences of the proposition's realization and, moreover, the overall outcome of the proposition's realization.

The considerations for a proposition's assessment may vary along a continuum of scope from logical consequences, as for intrinsic desires and aversions, to total outcome, as for extrinsic desires and aversions. Intermediate types of desire and aversion take account of the immediate, but not remote, causal consequences of a proposition's realization. For instance, an agent may be averse to going to the dentist, not because of its logical consequences, but because of its immediate causal consequences, for example, the pain of the procedures to be performed. Once the long-range causal consequences of going are also considered, including dental health, the aversion abates. The aversion springs from a type of assessment between a narrow assessment using logical consequences and a broad assessment using the entire outcome. To keep my theory simple, I put aside such intermediate types of desire, although they are common in everyday deliberations.

2.1.2. Basic Intrinsic Attitudes

Ends are objects of intrinsic desire, since otherwise desire for them would rest in part on their extrinsic features, and so on other things. But not all objects of intrinsic desire are ends. An intrinsic desire that p may arise because p's realization is a logical means to satisfaction of another intrinsic desire. A person, for instance, may have an intrinsic desire to be healthy and bright-eyed because of an intrinsic desire to be healthy. The conjunction of being healthy and bright-eyed may be desired just as a logical means to being healthy, and so not desired as an end.

Section 2.2.1's analysis of a proposition's utility in terms of an agent's goals assumes that the agent's goals are ends as well as objects of intrinsic desire. I take this to mean that they are objects of *basic* intrinsic desires. By a basic intrinsic desire I mean an intrinsic desire that is causally basic, and so not causally derivative from other intrinsic desires. I mean that the desire itself is causally basic, not that its object is causally basic, and so not causally derivative from the objects of other intrinsic desires. A basic intrinsic desire is a causally fundamental intrinsic desire, an "atomic" desire. Its object is desired for its own sake, not merely because of a desire for something else, even a logical consequence. An agent's basic intrinsic desires, BIDs, are the causes of all his other intrinsic desires, IDs, and their intensities. BIDs are not caused by any ID or combination of IDs. Instead they cause all other IDs. Of course, BIDs may vary from agent to agent, because the springs of desire may vary from agent to agent. One agent may be moved by altruism, whereas another is not. One agent's psychological makeup may differ from another's.

I do not provide an account of causation among intrinsic desires but make these assumptions about it. If we omit the special case of an agent without any desires at all, all agents have basic intrinsic desires. For no one desires anything unless she desires something as an end; and desiring something as an end requires having intrinsic desires and, moreover, causally basic intrinsic desires. Also, multiple causes of an intrinsic desire are possible. An ID that $a \ \& \ b \ \& \ c$ may be caused by the IDs that a, that b, and that c, and also by the IDs that a and that b & c. So there might not be a set of IDs that is *the* cause of an ID. Still, for simplicity, I assume that an ID has a unique cause composed of causally fundamental IDs. In my example the ID that b & c is not causally fundamental if it is itself caused by the IDs that b and that c. Instead, if the IDs that a, that b, and that c are BIDs, then their combination is the root cause of the ID that a & b & c. Evaluation of a proposition according to goals, as I present it, presumes that for any ID not itself a BID, there is a unique combination of BIDs causing it.

For a rational ideal agent, if a and b are each objects of intrinsic desire, then their conjunction is an object of intrinsic desire. But an agent should not evaluate a proposition with respect to intrinsic desires for a, b, and a & b. That involves double-counting. Once a and b are counted, all the value of a & b is already counted. To avoid double-counting intrinsic desires, this chapter's method of utility analysis uses only BIDs when evaluating a proposition. The method does not omit relevant considerations by failing to use nonbasic intrinsic desires. The method counts a nonbasic intrinsic desire by counting its causes, the intrinsic desires that generate it. If the intrinsic desires that a and that b cause the intrinsic desire that a & b, then counting the former counts the latter. Section 2.2.2 expounds this view more fully.

My method of intrinsic utility analysis must take account of intrinsic aversions, IVs, as well as intrinsic desires. They also influence the comprehensive utilities analyzed. To avoid double-counting without fear of omission, intrinsic utility analysis focuses on basic intrinsic aversions, BIVs, or causally fundamental intrinsic aversions. BIVs are analogous to BIDs. They cause all other intrinsic aversions.

Intrinsic attitudes of indifference, IFs, similarly arise from basic intrinsic attitudes of indifference, BIFs. For instance, BIFs toward two propositions may generate an IF toward the propositions' conjunction. In some cases IFs may arise from the cancellation of an ID and an IV. For instance, an intrinsic desire that p and an intrinsic aversion toward q may together generate intrinsic indifference toward p & q if the desire and aversion have equal strength.

In general, basic intrinsic desires, aversions, and attitudes of indifference are causally responsible for all other intrinsic attitudes. That is, basic intrinsic attitudes, BITs, form the foundation for intrinsic attitudes, ITs.

Intrinsic utility analysis need not attend to all intrinsic attitudes. For simplicity, it may put aside intrinsic attitudes of indifference and their relationships since they do not influence the comprehensive utilities analyzed. Because intrinsic attitudes of indifference are negligible in utility analysis, I generally ignore them. However, later I use the general attitude of intrinsic indifference as the zero point for a scale of intrinsic utility comprehending intrinsic desire, intrinsic aversion, and intrinsic indifference. I place intrinsic desires and intrinsic aversions on the scale according to their intensities, and use intrinsic indifference to separate them.

Simplification suggests defining intrinsic indifference in terms of intrinsic desire and aversion. It suggests taking intrinsic indifference as the absence of intrinsic desire and aversion. This reduction ignores a psychological distinction, however. Intrinsic indifference indicates a judgment, not the absence of judgments. Simplification also suggests taking intrinsic aversions as intrinsic desires that such and such not happen. This reduction similarly conflates a psychological distinction. An aversion is not the same as a desire that a negation hold. Someone cognitively deficient may have an intrinsic aversion but lack the concept of negation and so lack an intrinsic desire concerning the negation of the aversion's object. Someone inattentive or irrational may have an intrinsic aversion but fail to consider the negation of the aversion's object and so may also lack an intrinsic desire concerning the negation.

Because my objective is an account of intrinsic desires and aversions that serves as a foundation for decision rules for agents of all sorts – ideal and nonideal, rational and irrational – I have reason to preserve psychological distinctions between intrinsic attitudes that cognitive deficiencies, inattention, and irrationality reveal. But because I focus on rational ideal agents, it may seem useful to at least temporarily conflate attitudes that are functionally equivalent in the deliberations of such agents.

For comparison, consider an extrinsic aversion, the familiar type of aversion. An extrinsic aversion is not psychologically equivalent to a desire that the negation of the aversion's object hold. For irrational or nonideal agents, an extrinsic aversion may not generate an extrinsic desire that the negation of the aversion's object hold. For rational ideal agents, however, it does. Consequently, for such agents an extrinsic aversion functions in deliberations in the same way as an extrinsic desire that the negation of the aversion's object hold. A theory of their comprehensive utility assignments therefore may gloss over the distinction. May an account of their intrinsic utility assignments similarly gloss over analogous psychological distinctions between intrinsic attitudes?

Provided that the idealization for agents includes, along with rationality and cognitive power, the formation of intrinsic attitudes toward propositions appearing in intrinsic utility analyses, intrinsic difference may be reduced to the absence of intrinsic desire and aversion. After ruling out intrinsic desire and aversion, intrinsic indifference is the only intrinsic attitude remaining. This reduction does not yield an important simplification of intrinsic utility analysis, however. My version ignores intrinsic indifference except to establish a zero point for intrinsic utility. The reduction is not rewarding enough to make official.

A reduction of intrinsic aversion to intrinsic desire, or the reverse, seems possible for rational ideal agents because coherence may require each attitude to be accompanied by its counterpart. For example, if an agent has an intrinsic aversion toward p, coherence may require an intrinsic desire that $\sim p$.

Suppose coherence establishes a correlation between intrinsic desire and aversion. Should one attitude be reduced to the other? Such a reduction buries the coherence requirement that generates the correlation. To exhibit rationality's constraints on intrinsic attitudes, a theory of intrinsic utility should forgo the reduction. The theory may put some rationality constraints into background idealizations about agents, but it should not hide constraints applying to its main topic.

Also, intrinsic utility analysis rests on causally basic intrinsic attitudes. To preserve its foundation, a reduction of attitudes must preserve the BITs. If a BID causes an IV, the IV should be reduced to the BID; and if a BIV causes an ID, the ID should be reduced to the BIV. A reduction of attitudes should reduce the effect not the cause. Consequently, if the causal connections between intrinsic desires and aversions run in both directions, a reduction of attitudes may not uniformly reduce intrinsic aversions to intrinsic desires, or the reverse. Hence no useful simplification results.

For the foregoing reasons, a theory of intrinsic utility should not use a correlation of intrinsic desire and intrinsic aversion to reduce one to the other. Furthermore, first impressions notwithstanding, coherence does not create a correlation of intrinsic desire and intrinsic aversion in rational ideal agents. Although in a rational ideal agent, an extrinsic attitude toward a proposition is accompanied by the opposite extrinsic attitude toward the proposition's negation, the correlation does not extend to intrinsic attitudes. Coherence requirements for intrinsic desires and intrinsic aversions do not replicate coherence requirements for extrinsic desires and extrinsic aversions. Extrinsic attitudes evaluate a proposition's causal consequences, including frustration of desires. However, intrinsic attitudes evaluate a proposition's logical consequences, which generally exclude frustration of desires. The more narrow scope of intrinsic attitudes weakens coherence requirements for them.

Suppose that a rational ideal agent has an ID that p. Must he have an IV toward $\sim p$? Frustration of desire is not a logical consequence of $\sim p$. It is an extrinsic feature of $\sim p$ that its realization thwarts an intrinsic desire. Frustration follows only given $\sim p$'s realization and a premiss stating the agent's intrinsic desire that p. Knowledge of an agent's intrinsic attitudes is a posteriori knowledge, and goes beyond knowledge of a proposition's logical consequences. An aversion's dependence on such information indicates an extrinsic aversion. Although the logical consequences of $\sim p$ may independently prompt an intrinsic aversion, they need not. Coherence requires that if an agent has an intrinsic desire that p, he lacks an intrinsic desire that $\sim p$. Under the idealizations, this leaves open two possibilities concerning $\sim p$: intrinsic aversion toward $\sim p$ and intrinsic indifference toward $\sim p$. The negation's logical consequences may prompt intrinsic indifference rather than aversion. Intrinsic attitudes toward p and $\sim$p do not stem from an evaluation of the logical consequences of just one proposition. They rest on evaluations of each proposition's logical consequences.

To illustrate, imagine that a rational ideal agent considers pleasure, pain, and the absence of both. She forms an intrinsic desire for pleasure and an intrinsic aversion to pain. What follows about her intrinsic attitude to the absence of pleasure? Coherence does not demand an intrinsic aversion to pleasure's absence. It leaves open the possibility of intrinsic indifference to pleasure's absence. Pleasure's absence does not entail pain so the agent may have intrinsic indifference toward it. She may have intrinsic indifference to pleasure's absence, pain's absence, and the absence of both pleasure and pain.

Next, suppose that Adam has basic intrinsic desires to be healthy and to be wise, and no other basic intrinsic desires or aversions. Let Ha and Wa abbreviate the objects of his BIDs. The realization of $\sim Ha$ does not entail anything about Wa. Also, it does not entail the realization of any basic intrinsic aversion; Adam has no basic intrinsic aversions. It does entail frustration of the basic intrinsic desire that Ha. But this grounds only an extrinsic aversion toward $\sim Ha$, not an intrinsic aversion. So Adam's intrinsic attitude toward $\sim Ha$ may be intrinsic indifference, not intrinsic aversion. If typical, Adam has intrinsic indifference toward $\sim Ha$ and extrinsic aversion toward $\sim Ha$. He may have no intrinsic aversions at all.

I have argued against a coherence requirement generating a correlation of intrinsic desires and aversions in rational ideal agents. The absence of such a coherence requirement leads to some conclusions about the causal relationship between intrinsic desires and aversions in rational ideal agents. In such agents causal relationships between intrinsic attitudes follow reasons – in particular, coherence requirements. Without a coherence requirement linking intrinsic desires and aversions, an intrinsic desire or aversion is not a sufficient reason for an opposite attitude. Hence an intrinsic desire or aversion does not by itself cause an opposite attitude. If the opposite arises, another factor participates in its creation. That is, the desire or aversion is not a sufficient cause of the opposite attitude.[4]

Suppose that an intrinsic desire that $\sim p$ and an intrinsic aversion toward p coexist. Typically, they arise independently. The desire stems from an evaluation of $\sim p$'s logical consequences, and the aversion stems from an evaluation of p's logical consequences. Sometimes, however, the intrinsic aversion may play a role in the desire's production. For example, a second-order intrinsic desire and an intrinsic aversion may causally interact to produce an intrinsic desire for nonrealization of the intrinsic aversion. Suppose a rational ideal agent has a general second-order intrinsic desire for the nonrealization of intrinsic aversions (an extrinsic desire for their nonrealization would be more typical). Then an intrinsic aversion may cause an intrinsic desire for nonrealization of the aversion. It may do this by causing a belief that the intrinsic aversion exists, which together with the second-order desire, causes an intrinsic desire for nonrealization of the aversion. Still, the intrinsic aversion is not a sufficient cause of the intrinsic desire for its nonrealization; it operates in conjunction with the second-order desire.

Intrinsic desires and intrinsic aversions are causally independent in the sense that no intrinsic desire is a sufficient cause of an intrinsic aversion, and no intrinsic aversion is a sufficient cause of an intrinsic

---

[4] A sufficient cause invariably produces its effect. The relation of causal sufficiency obtains between event types, not event tokens. An event token may produce an effect token, although the corresponding event type is not a sufficient cause of the corresponding effect type.

desire. Intrinsic desires and intrinsic aversions inhabit causally separate spheres in this sense. Similarly, for rational ideal agents attitudes of intrinsic indifference inhabit a sphere causally separate from the spheres of intrinsic desires and aversions. Suppose that an agent has neither an intrinsic desire nor an intrinsic aversion toward a possible world. Then he has intrinsic indifference toward the world. It arises from his evaluation of the world's features and may not be caused by his intrinsic desires and aversions. Suppose, however, that his intrinsic desires and aversions to features of the world cancel each other and thereby cause his indifference to the world. Still, neither the desires by themselves nor the aversions by themselves suffice to cause his indifference. Both participate in generating his intrinsic indifference toward the world. Events in one sphere are not enough to cause events in another sphere. Each sphere is causally independent of every other sphere. As a foundation for intrinsic utility analysis, Section 2.1.5 presents some principles of coherence that assume this causal independence.[5]

2.1.3. Conditional Intrinsic Attitudes

There are also intrinsic attitudes given conditions, for instance, a desire for a long life given that one is healthy. I distinguish types of conditional intrinsic attitudes according to the manner in which the condition is given. For instance, an intrinsic desire for a long life if one *is* healthy differs from an intrinsic desire for a long life if one *were* healthy. The manner of supposition appropriate for a condition depends on the condition and context. In deliberation, indicative supposition is appropriate for a state; subjunctive supposition is appropriate for an option. As Section 4.2 explains, indicative supposition suits evidential considerations and subjunctive supposition suits matters one controls.

Basic intrinsic attitudes (BITs) generally do not change when taken with respect to conditions that carry new information, since those atti-

---

[5] Because intrinsic desires (IDs) and intrinsic aversions (IVs) are causally independent, it is possible for a rational ideal agent to have a BID that p and a BIV that $\sim p$. There may be independent grounds for each attitude. Each may arise because of the logical consequences of its object. It need not be that one attitude causes the other and makes it nonbasic. As a result, there may be rational ideal agents for whom every possible world involves the realization of some BIT.

tudes are generally independent of changes in information. In any case, Section 1.3.2's idealization that goals are stable ensures that BITs are stable given conditions entertained in utility analysis – in particular, given options. My principles of intrinsic utility analysis take advantage of this idealization. They put aside, for instance, cases in which an option, if it were realized, would cause changes in BITs.[6]

The idealization that conditions leave BITs unchanged still allows a condition to affect intrinsic attitudes. It can do this by specifying or precluding partial realization of an intrinsic attitude's object. For instance, one may intrinsically desire that a & b, but not either that a or that b. Then b is intrinsically desired given a, but not otherwise. The conditional desire focuses on the common component of the b-worlds among the a-worlds. It does not focus on the common component of all b-worlds, as the intrinsic attitude toward b does. A conditional intrinsic desire for a proposition's realization thus may differ in intensity from a nonconditional intrinsic desire for the proposition's realization, even though the proposition's logical consequences are invariant world to world and BITs are immune from revision when taken with respect to a condition.

Conditions do not affect intrinsic attitudes toward possible worlds, however. Because a world is maximal, it specifies for every BIT whether that BIT is realized. Then, because BITs do not change given a condition, neither does an intrinsic attitude toward the world.

2.1.4. Complete Sets of Basic Intrinsic Attitudes

A set of basic intrinsic attitudes of a person is *complete* if and only if all the person's other intrinsic attitudes are generated by them. I put aside attitudes of intrinsic indifference, however, and take a set of basic intrinsic attitudes to be complete if it generates all other intrinsic desires and aversions. Intrinsic desires and aversions are the intrinsic attitudes of interest.

Finding a complete set of basic intrinsic desires and aversions for a person is similar to finding a complete set of axioms for the truths of a scientific theory. For an ideal agent it amounts to finding elements for the generation of a set of possible worlds restricted to matters

---

[6] Suppose that the condition entertained is that BITs change. The intrinsic attitude toward a proposition, given that condition, still depends on current BITs. Current BITs assess the situation specified by the proposition and new BITs.

of concern. This is not an easy task; it is hard to find a reduction of intrinsic attitudes to BITs. In realistic cases we are usually not able to list BITs thoroughly and accurately.

To illustrate typical problems, consider a case where desires concerning pleasure are the only intrinsic desires. Suppose that t_1 and t_2 are adjacent temporal intervals, and an agent wants pleasure of degree n during each interval. Furthermore, suppose he wants a constant degree of pleasure during both intervals. Imagine that all three desires are intrinsic, and there are no other causally independent intrinsic desires. Which of these three intrinsic desires, if any, is basic? Satisfaction of the desires for pleasure of degree n during the intervals entails satisfaction of the desire for constancy of pleasure. Do those desires cause the desire for constancy? Maybe not. The desire for constancy is not required by coherence. So the desire for constancy of pleasure may be basic despite its satisfaction's being entailed by satisfaction of the desires for the two pleasures. All three desires may be basic. They may be causally independent although their objects are not logically independent.

The desires for the two pleasures need not be basic, however. Perhaps the desires for pleasure of degree n during the intervals are caused by other intrinsic desires, say, intrinsic desires for pleasures of a certain amount, where an amount is defined as the product of the pleasure's intensity and duration. Satisfaction of desires for amounts of pleasure does not entail satisfaction of the desire for constancy of pleasure, because an amount of pleasure may be produced by many pleasures of various intensities and durations. These desires for amounts of pleasure clearly do not cause the desire for constancy of pleasure. So the desire for constancy may be basic, whereas the desires for pleasures of degree n are not.

Another possibility is that an intrinsic desire for pleasure of any degree any time might cause the desires for pleasure of degree n during t_1 and during t_2. The general desire for pleasure might cause the desires for these two pleasures. Still another possibility is that the objects of BIDs for pleasure are fine-grained, perhaps propositions with concrete possibilia as components. There may be an intrinsic desire for *this* pleasure and a desire for *that* one afterward. Perhaps the desires for pleasure of degree n are caused by basic desires for particular pleasures that realize those desires for pleasure in the abstract. Finally, it is possible that the causally basic desires' objects are more global than pleasures of certain degrees during the intervals t_1 and t_2. Their objects might be pleasure profiles for the interval composed of t_1 and t_2. These desires may cause both the desires for the two pleasures of degree n and the desire for constancy of pleasure. It is difficult to investigate all the possibilities and find the basic intrinsic desires.

One complication is that two factors may defeat an ID's bid to be a BID. Other IDs may either cause its existence or influence its intensity. Suppose that there is an ID for pleasure and an independently existing ID for pleasure of degree n. In a rational ideal agent the ID for pleasure boosts the intensity of the ID for pleasure of degree n because satisfaction of the latter entails satisfaction of the former. The ID for pleasure of degree n is therefore not a BID.

Despite the difficulty of a causal account of intrinsic attitudes in terms of basic intrinsic attitudes, ideal agents, given their cognitive powers and self-awareness, know their basic intrinsic attitudes (BITs). Utility analysis in terms of BITs, to be presented later, is available to these agents. In examples of these analyses, I generally assume that the causal breakdown of intrinsic attitudes into basic intrinsic desires and aversions does not yield BITs for worlds, but for less global propositions, so that BITs cause intrinsic attitudes toward worlds.

2.1.5. Principles Governing Intrinsic Attitudes

This section advances some principles for intrinsic desires, in particular, laws of coherence. Laws of coherence govern intrinsic desires in the way they govern extrinsic desires and beliefs. People may sometimes violate, and excusably violate, these laws, but rational ideal agents observe them. Although I focus on laws for intrinsic desires (IDs), there are similar laws for intrinsic aversions (IVs).

One law of coherence prohibits having an ID that p simultaneously with an ID that $\sim p$. Another law of coherence says that if p and q are logically equivalent, then p is intrinsically desired if and only if q is. A law concerning conjunctions requires that if there are IDs that p and that q, then there is an ID that p & q. The intrinsic desires that p and that q need not cause the intrinsic desire that p & q. Causation might go in the reverse direction. However, if the intrinsic desires that p and that q are basic, then I claim that they cause an ID that p & q in a rational ideal agent because p & q is the conjunction of their objects. I do not claim that a BID that p and a BID that q cause an ID that p & q in a rational ideal agent because p and q together entail p & q. The desires for the two pleasures in the preceding example need not cause

57

the desire for constancy of pleasure, although the two pleasures entail constancy. A rational ideal agent may not intrinsically desire all the logical consequences of what he intrinsically desires. An intrinsic desire may depend on just some of its object's logical consequences; other logical consequences may be matters of indifference. A rational ideal agent may not even intrinsically desire all the logical consequences of a BID's object. For instance, a rational ideal agent may have a BID that p & q, but lack an ID that p even though p is a logical consequence of p & q.

Some candidate laws of coherence are passed over. The objects of atomic desires need not be logical atoms. The objects of BIDs need not be logically independent atomistically (so that no two objects share logical atoms). There may be BIDs that p & q and that q & r. Also, the objects of BIDs need not be strongly logically independent (so that no set's satisfaction and frustration pattern entails any nonmember's satisfaction or frustration). There may be families of BIDs with objects that are mutually exclusive. For instance, there may be a BID for pleasure of degree n at t, and also a BID for pleasure of degree $n + 1$ at t. Here satisfaction of one BID entails frustration of the other. The objects of BIDs need not be even weakly logical independent (so that no set's satisfaction entails any nonmember's satisfaction). In the case of the two pleasures, for example, the intrinsic desire for each pleasure and the intrinsic desire for constancy may be basic, although satisfaction of the desires for the two pleasures entails satisfaction of the desire for constancy.[7]

Because of the logical relations among objects of BIDs, it may be impossible to satisfy all BIDs. Similarly, it may be impossible to realize all BIVs. Also, the nonrealization of a BIT or group of BITs need not entail anything about the realization of other BITs. As a result, with respect to an agent's set of BITs, there may be a possible world in which no BIT is realized (putting aside IFs, as is my custom).

Laws of coherence may address causes of attitudes as well as their objects. To be sure, causal derivation of attitudes is distinct from logical derivation of their objects, and intrinsic desires and aversions are classified as basic, not their objects. Still, logical relations among the objects of intrinsic desires and aversions have implications for causal relations among the desires and aversions if agents are ideal. In a rational ideal

---

[7] This stand on independence revises the view expressed in Weirich (1987).

58

agent intrinsic attitudes, which assess logical consequences, respond to logical relations among propositions. Certain laws of coherence with overt causal implications I call *causal principles*. I advance three for intrinsic desires.

The first causal principle describes a way BIDs generate IDs. It generalizes an earlier claim that a pair of BIDs causes an ID that the conjunction of their objects hold.

A set of multiple BIDs causes an ID for the realization of the set of BIDs.

This principle holds for a set of any number of BIDs. Because the realization of a set of BIDs is logically equivalent to the realization of the conjunction of the BITs' objects, the set of BIDs also generates an ID for that conjunction's realization. Under the assumption that an intrinsic attitude has a unique cause, the set of BIDs is the unique cause of that ID.

The second causal principle states that BIDs are atomic causes of IDs, or fine-grained reasons for IDs.

No BID's object entails a distinct BID's object.

According to this principle, a BID that p rules out a BID that p & q since p & q entails p. The object of one BID cannot entail the object of another because given the entailment they cannot possess the causal independence required of BIDs. A BID influences the intrinsic attitude toward propositions entailing its object. The first principle implies a related point. It prohibits a BID that p & q given a BID that p and a BID that q since it claims that the simple BIDs would cause the complex BID, and a BID is causally independent. The second principle, however, implies even stronger prohibitions concerning BIDs. Given only a BID that p it prohibits a BID that p & q and also a BID that $p \lor q$.

The third causal principle establishes a connection between the realization of an ID and the realization of BIDs causing the ID. It is for nonbasic IDs.

An ID's realization also realizes some BIDs causing the ID.

For example, suppose that a BID that p and a BID that q cause an ID that p & q. Then the ID's realization achieves the realization of both BIDs causing the ID. Also, suppose that the same BIDs cause an ID that $p \vee q$. Then that ID's realization achieves the realization of one or both of the BIDs. The principle holds because an ID assesses logical consequences. Hence each of its realizations must achieve some goals prompting it. Otherwise the ID would be insufficiently grounded in logical consequences.[8]

The three causal principles are compatible. Imagine a case with an ID that p & q, a BID that p, and a BID that q, and no other IDs. Suppose the BIDs that p and that q together cause the ID that p & q. The first principle is satisfied given this pattern of causation. The second principle is satisfied, assuming that p and q are logically independent. And the third principle is satisfied because realizing the ID that p & q realizes the BIDs that cause it.

Some initially appealing causal principles do not pass muster. Take this principle: an intrinsic desire's object entails the objects of the intrinsic desires that cause it. It claims, for instance, that if the intrinsic desire that p is caused by the intrinsic desires that q and that r, then p entails q and r. This happens if p is the conjunction of q and r. But it does not happen if p is the disjunction of q and r. It is also false that if a proposition entails another, then an intrinsic desire for the latter causes an intrinsic desire for the former. For p & q entails p, but an ID that p need not cause an ID that p & q. Perhaps q is an object of over-whelming intrinsic aversion.

Next, consider this principle: no BID's object entails the object of another ID. It assumes that for a rational ideal agent, an ID for a distinct logical consequence of another ID's object is part of the latter ID's cause. Then it observes that a BID is causally independent and so cannot have such a cause. For example, an ID that p is part of the cause of an ID that p & q if both desires arise, so an ID that p & q cannot be a BID given an ID that p. However, the principle's assumption does not hold in all cases. An ID that $p \vee q$ may arise from a BID that p and a BID that q. The object of each BID entails $p \vee q$. But the ID that $p \vee q$ does not cause the intrinsic attitudes toward p and toward q. In this case the BID that p and the ID that $p \vee q$ form a counterexample to the principle.

The three causal principles adopted for a rational ideal agent's intrinsic desires also apply, mutatis mutandis, to the larger realm of a rational ideal agent's intrinsic attitudes. The general principles are:

---

[8] A complementary principle, but one less important for utility analysis, states that a BID does not cause an ID unless it is realized in some realization of the ID.

(CP1) A set of multiple BITs causes an IT toward the realization of the set of BITs.

(CP2) No BIT's object entails a distinct BIT's object.

(CP3) An IT's realization also realizes some BITs causing the IT.

CP1 applies to sets of BITs mixing BIDs and BIVs. An IT toward a mixed set's realization arises from the BITs in the set, just as an ID arises from a set of BIDs. CP2 presumes that intrinsic indifference has been set aside. For a rational ideal agent might have a basic intrinsic desire that p & q and basic intrinsic indifference toward q. Putting aside such cases by putting aside intrinsic indifference, the principle survives. Whether the remaining BITs are uniformly BIDs or BIVs, or mixtures of BIDs and BIVs, their causal independence backs CP2. Section 2.1.2's claim that intrinsic desire, indifference, and aversion form independent causal spheres supports CP3, which concerns nonbasic ITs. It rules out a case in which a BID that p by itself causes an IV that $\sim p$, a case that contravenes the principle because in it $\sim p$'s realization does not realize any of its causes.

The next section's argument for its method of calculating the intrinsic utility of a possible world draws on the three causal principles for intrinsic attitudes. They furnish key premises of the argument. This section's other points about coherence, and causal and logical relationships, are intended only to clarify intrinsic attitudes. The next section's argument does not use them.

2.2. INTRINSIC UTILITIES OF WORLDS

Intrinsic utility rests on intrinsic attitudes. After introducing intrinsic utility, this section distinguishes it from comprehensive utility – that is, utility as standardly conceived in decision theory – and shows how to compute the intrinsic utility of a possible world.

2.2.1. Intrinsic Utility

Utility, as commonly understood in decision theory, is rational degree of desire. It registers both extrinsic and intrinsic desire. Because it is inclusive, as is extrinsic desire, it might be called "extrinsic utility." I take intrinsic utility as rational degree of intrinsic desire. It puts aside extrinsic desire and focuses exclusively on intrinsic desire. Although it resembles utility *tout court*, it imposes a restriction on the scope of a proposition's assessment. A proposition's utility is a rational degree of desire considering the outcome if the proposition were true. This outcome is the complete outcome of the proposition's realization, including all causal consequences. Utility *tout court* is comprehensive, and I often call it *comprehensive utility* to emphasize this feature. On the other hand, a proposition's intrinsic utility is a rational degree of desire considering only the proposition's logical consequences. The causal consequences of the proposition's realization, and other parts of its outcome, are excluded because the proposition does not entail them.

When defining intrinsic utilities as rational degrees of intrinsic desire, I take degrees of intrinsic desire to measure intrinsic aversions as well as desires. Also, I take them to measure intrinsic indifference; all intrinsic attitudes are covered. The intrinsic utility of an object of intrinsic desire is positive. The intrinsic utility of an object of intrinsic indifference is zero. The intrinsic utility of an object of intrinsic aversion is negative, the magnitude or absolute value of the negative quantity increasing with the intensity of the intrinsic aversion.

Both comprehensive and intrinsic utility are subjective, dependent on an agent's desires, but intrinsic utility is independent of information because logical consequences are independent of information. Despite being subjective, comprehensive and intrinsic utility have structure because they meet rationality constraints. Because intrinsic utilities are *rational* degrees of intrinsic desire, an ideal agent's intrinsic utility function conforms with the laws of coherence in the previous section.[9]

To bring out the difference between comprehensive and intrinsic utility, I use their different types of assessment to explain their

---

[9] Binmore (1998: 362–3) distinguishes between direct and indirect utility functions. Direct utility functions apply to final ends, whereas indirect utility functions apply to means. An act is a means to its consequences or outcomes, which are taken to be final ends. Accordingly, direct utility functions apply to consequences or outcomes, whereas indirect utility functions apply to acts. Indirect utility functions are sensitive to changes in information, but direct utility functions are not. This distinction between utility functions is similar to the distinction between intrinsic utility and comprehensive utility. Intrinsic utility is roughly direct utility, and comprehensive utility is roughly indirect utility. The main differences are that (1) for theoretical simplicity comprehensive utility applies to all propositions, those expressing acts and those expressing consequences and outcomes, and (2) an act's consequences and outcomes on my view may not be realizations of final ends but may instead comprehend realizations of final ends in a way that makes them means of attaining final ends.

different roles in decision principles. Comprehensive utility is the main resource of decision principles because it assesses motivation. Intrinsic utility, in contrast, assesses satisfaction. It assesses a proposition according to the satisfaction the proposition (logically) guarantees – that is, the level of realization of basic intrinsic attitudes the proposition guarantees. Here "satisfaction" is a technical version of the ordinary term. I suppose that amounts of satisfaction are possible and, in particular, that negative amounts of satisfaction are possible. I do not introduce the technical term in detail because I do not use it to define intrinsic utility but only to provide a rough characterization of intrinsic utility that contrasts it with comprehensive utility.

An agent may be motivated to undertake a gamble that might not result in the realization of any basic intrinsic attitude. His motivation may spring from the probability he assigns to the gamble's resulting in a BIT's realization. Assessments of motivation are dependent on information, whereas assessments of guaranteed satisfaction are independent of information. Comprehensive utility reacts to changes in the subjective probabilities of realizations of BITs as information changes. It arises from BITs but takes account of information. When information is incomplete, comprehensive utility's sensitivity to information makes it a better guide to decisions than intrinsic utility. My decision theory introduces intrinsic utility only to calculate comprehensive utilities.

The difference between intrinsic utility and comprehensive utility is sharp. The intrinsic utility that Smith assigns to the proposition that she receives a check for $1 million may be low because the proposition's logical consequences do not ensure much satisfaction. The possibility of forgery, hyperinflation, and the collapse of the monetary system weaken the guarantee. However, the comprehensive utility she assigns to the proposition may be high because she thinks it likely that receiving the check will enrich her life. Similarly, the intrinsic utility that Jones assigns to the proposition that he mails a letter may be low because logic does not guarantee delivery. Nonetheless, the comprehensive utility he assigns to the proposition may be high because he thinks the postal service is reliable.

The features of intrinsic desire direct intrinsic utility's assessment of propositions. Intrinsic desire focuses on a proposition's logical consequences and evaluates the amount of satisfaction they contribute in worlds realizing the proposition. As a result, intrinsic utility evaluates the contribution a proposition's logical consequences make to satisfaction. Because it attends to that contribution, its assessment of a proposition need not agree with comprehensive utility's assessment of the proposition, even given full information about the outcome of the proposition's realization. For example, the intrinsic utility of a BID's realization may differ from the comprehensive utility of the BID's realization, given full information, because the intrinsic utility assesses contribution to satisfaction, whereas the comprehensive utility assesses total satisfaction. Imagine a case with a BID that p and a BID that q. The first may be realized with and without the second's realization. Consequently, p-worlds may yield different levels of satisfaction. The intrinsic utility of p is the amount of satisfaction p's realization adds to each p-world, in particular, if it exists, a p-world that realizes no other BITs.

Because an intrinsic utility is quantitative, a proposition has an intrinsic utility only when its realization guarantees a definite amount of satisfaction. A proposition qualifies if its realization contributes the same amount of satisfaction in every world realizing it. However, not every proposition qualifies. A proposition's contribution may be variable. Then its realization guarantees no precise amount of satisfaction. Consequently, it lacks an intrinsic utility. For example, suppose an agent has an ID for pleasure. Various circumstances may accompany pleasure, and pleasure itself may come in various forms. The ID surveys the amounts of satisfaction attributable to pleasure in each of its realizations. If those amounts vary, then pleasure's intrinsic utility does not exist. Next, suppose that p and q are logically incompatible, and that a BID that p and a BID that q cause an ID that $p \vee q$. The disjunction promises the amount of satisfaction p generates or the amount of satisfaction q generates or the amount of satisfaction p and q generate together. If those amounts differ, it guarantees no precise amount of satisfaction. Then even an ideal agent may not assign an intrinsic utility to $p \vee q$. In fact, given a BIV toward q, the agent may not even form an intrinsic attitude toward $p \vee q$ – not desire, indifference, or aversion. The disjunction's realization may not guarantee satisfaction, equanimity, or frustration, much less an amount of satisfaction (in the technical sense that permits negative amounts).

A proposition's logical consequences are the common component of worlds that realize it. The intrinsic utility of a proposition p, $IU(p)$, exploits this fact. It assesses p's logical consequences by reviewing the set of p-worlds. $IU(p)$ is the amount of satisfaction that p's logical consequences contribute in any world realizing p. Although p's logical consequences are constant from one p-world to another, their contribution may vary with the p-world in which they occur. For example, suppose p is $q \vee r$. The logical consequences of $q \vee r$ may contribute different amounts of satisfaction in p-worlds, depending on whether they are q-worlds or r-worlds. Such variation may block the formation of p's intrinsic utility. $IU(p)$ does not exist if p does not contribute a definite amount of satisfaction in every p-world.

The comprehensive utility of a proposition p, $U(p)$, surveys the set of p-worlds, too. Causal variants of $U(p)$ track p's causal consequences from p-world to p-world. These consequences may vary. They are evaluated by calculating the probability-weighted average of the amount of satisfaction they generate in p-worlds. $U(p)$ itself, being comprehensive, evaluates entire outcomes, not just causal consequences. It calculates the probability-weighted average amount of satisfaction in p-worlds (see Chapter 3). The different scope and method of evaluation accounts for $U(p)$'s difference from $IU(p)$.

A BIT's realization contributes the same amount of satisfaction in every world realizing the BIT. Its causal independence ensures this. A possible world's realization contributes the same amount of satisfaction in every world realizing it, because there is only one such world. Realizations of BITs and possible worlds guarantee definite amounts of satisfaction. An ideal agent therefore assigns intrinsic utilities to objects of BITs and possible worlds. Intrinsic utility analysis involves intrinsic utilities of possible worlds and objects of BITs. It traffics only in propositions that have intrinsic utilities.

To bring out further the difference between comprehensive utility U and intrinsic utility IU, I compare their assessment of a BIT's object. Let BIT_j stand for the object of the jth BIT in an arbitrary order. $IU(BIT_j)$ may not equal $U(BIT_j)$ if the realization of BIT_j causes the realization of other BITs. This is a possibility even though BITs are causally independent since the desires' causal independence does not entail their objects' causal independence.

Also, I compare comprehensive and intrinsic utility's assessment of chances. It is helpful to do this now, although chance does not play a role in intrinsic utility analysis until Chapter 5, because the contrast is so striking. It dramatizes the difference between the two types of utility.

A mere chance of satisfaction of an intrinsic desire is not intrinsically desirable, even if empirical information makes the chance very high. For instance, suppose that an agent has an intrinsic desire that p. A chance that realization of $p \vee q$ results in realization of p does not suffice to generate an intrinsic desire that $p \vee q$. Even a 100 percent chance that realization of $p \lor q$ brings realization of p is not enough to generate an intrinsic desire that $p \lor q$. Intrinsic desire goes by logical consequences, not sure accompaniments, and the disjunction does not entail p. Intrinsic desires, and intrinsic utilities, are independent of information because they assess logical consequences of propositions. Intrinsic utility's concentration on logical consequences also makes it insensitive to subjective probabilities, which are relative to information.

Extrinsic desires, on which comprehensive utility or U rests, take account of information. Extrinsic desires stem from intrinsic desires and information. U applies to propositions and attempts to evaluate complete outcomes and so all consequences, causal and logical alike. Given uncertainty, it evaluates subjective chances for outcomes.

$U(p)$ goes by p's creation of subjective chances for realization of BITs. If p's realization creates a chance for satisfaction of a BID, the utility of that chance is a reason to desire that p hold, and $U(p)$ is the combined strength of all such reasons for and against p's realization. $U(p)$ treats p as a lottery over BITs, a lottery that obtains if p obtains. $U(p)$ is the sum of the utilities of the subjective chances of realization of BITs created by p's realization (see Section 5.2.3). $IU(p)$, in contrast, restricts its attention to p's logical consequences. Chances p's realization creates are not logical consequences but dependent on the agent's beliefs. $IU(p)$ measures the level of satisfaction from BITs that p logically guarantees.[10]

Intrinsic and comprehensive utilities have many features in common. They both evaluate propositions in terms of rational desire. They differ chiefly in the range of desires they consider, the style of their evaluations, and their dependence on empirical information.[11]

---

[10] Possibilities of realization of BITs may be entailed by p. But the chances for realization of BITs on which $U(p)$ depends are not entailed by p, even though they are certain given p's realization. The size of the chances depends on information, and so the chances are not entailed by p alone. Realization of a lottery over realizations of BITs creates certainty of chances for realizations of BITs but entailment of only possibilities of realizations of BITs. The chances are certain but not entailed by the proposition that the lottery is realized, because the (subjective) chances' sizes depend on the agent's information. The possibilities of the chances (whatever those chances are) are entailed by the proposition, but not the chances themselves. Intrinsic utility responds to logical guarantees of BITs' realizations not mere possibilities. $IU(p)$ does not register the intrinsic value of the chances for realizations of BITs because they are not entailed.

[11] Are there alternative accounts of the distinction between comprehensive and intrinsic utility? One approach is to take U and IU as the same function applied to dif-

66

2.2.2. A World's Intrinsic Utility

I take a possible world to be a maximal consistent proposition (limited to matters of concern). Because a world is a proposition, it has an intrinsic utility and a comprehensive utility. Intrinsic utility analysis computes utilities from intrinsic utilities. This section uses it to compute a world's intrinsic utility from intrinsic utilities of realizing basic intrinsic attitudes. Section 2.3 uses it to compute a world's comprehensive utility from intrinsic utilities of realizing BITs.

To start, consider the intrinsic utilities on which my calculations rest, the intrinsic utilities of realizing BITs. The intrinsic utility of satisfying a basic intrinsic desire is a very simple intrinsic utility. It is just the strength of the intrinsic desire for the BID's satisfaction – that is, the strength of the desire for the BID's satisfaction attending to only its logical consequences and putting aside its causal consequences and other parts of its complete outcome. Analogous remarks apply to basic intrinsic aversions. Because basic intrinsic indifference has no role, except as a zero point, the building blocks of utility calculations are simple intrinsic utilities of realizations of BIDs and BIVs.

ferent objects. One might propose thinking of them as the same rational degree of desire function over different objects, one broad (the entire outcome of a proposition's realization) and the other narrow (a proposition's logical consequences only). In place of $U(p)$ and $IU(p)$ as symbols for the utility and intrinsic utility of p, one might substitute $D(O[p])$ and $D(L[p])$, where D indicates rational degree of desire, $O[p]$ stands for the outcome of p (including causal consequences), and $L[p]$ stands for the logical consequences of p. This approach needs a specification of a suitable type of desire. Intrinsic desire focuses on $O[p]$'s logical consequences, which defeats the characterization of U, if, because of uncertainty, $O[p]$ does not fully specify p's outcome. Extrinsic desire reviews the causal consequences of $L[p]$, which defeats the characterization of IU because those causal consequences are the same as p's.

In any case, the alternative account yields an awkward utility theory. To facilitate mixing types of utility analysis, it is better to let the types of utility have the same objects and alter the scope of their evaluation. It is better to take comprehensive utility and intrinsic utility as different functions over the same set of propositions, not the same function over different sets of propositions or other objects. Using the same objects for comprehensive and intrinsic utility facilitates the formulation of laws connecting the two types of utility.

For the sake of uniformity in objects of utility, I also put aside a characterization of intrinsic utility that applies it to properties. It starts by taking the objects of intrinsic desire to be properties instead of propositions. The strength of one's intrinsic desire for a property on this view equals the strength of one's intrinsic desire that the actual world instantiate that property on my view. The intrinsic utility of a property is then the intrinsic desirability of the actual world's instantiating the property. Focusing on properties, intrinsic utility puts aside causal consequences for

Next, consider a possible world's intrinsic utility. I assume that, for each BIT and possible world, whether the BIT is realized in the possible world is a determinate matter. A possible world's features depend on the agent's concerns, in particular, the objects of his BITs, even if these are vague. If the object of a BIT is vague, whether it is realized in the world is still a determinate matter, because worlds, being propositions, are vague in ways that match the objects of BITs. Given my assumption, the intrinsic utilities of the objects of BITs may be used to calculate the intrinsic utility of a possible world. The intrinsic utility of the world is the sum of the intrinsic utilities of the objects of the BITs realized there. Notice that the BITs realized are the same as the BITs whose realization is entailed by the world because, being a maximal consistent proposition, a world entails each of its features.

Here is an example. Adam has two basic intrinsic attitudes, intrinsic desires to be healthy and to be wise. Possible worlds trimmed of irrelevant detail are:

w_1	Ha & Wa
w_2	Ha & $\sim Wa$
w_3	$\sim Ha$ & Wa
w_4	$\sim Ha$ & $\sim Wa$

In order for the intrinsic utilities of these worlds to be obtainable by addition of intrinsic utilities of BITs' objects, it must be the case that $IU(w_1) - IU(w_2) = IU(w_3) - IU(w_4)$. These differences are in fact equal. Suppose the second difference were bigger than the first. Then some complementarity between health and wisdom would make wisdom more intrinsically desirable given health than given illness. In an ideal agent there must be another BIT that causes this effect of health on the intrinsic desirability of wisdom – say, a BID for exercising wisdom.

---

logical consequences, because only propositions, not properties, have causal consequences. The effect is equivalent to assessing the proposition corresponding to the property's instantiation in terms of its logical rather than its causal consequences.

A variant of this account of intrinsic utility takes propositions as the objects of intrinsic utility but takes them as properties of worlds. Another variant distinguishes between a desire that p and a desire for p's holding. It takes the syntactic distinction to mark the distinction between propositions and properties as objects of desire and uses that-clauses to express the objects of comprehensive utility and gerundive nominalizations to express the objects of intrinsic utility.

The assumption that the basic intrinsic desires for health and for wisdom are the only BITs eliminates this possibility.[12]

As I shall argue, because of the nature of an ideal agent's BITs, the intrinsic utility of a world w is the sum of the intrinsic utilities of the objects of the BITs realized there.[13] Using Σ_j to stand for summation with respect to the index j, my principle stated compactly is:

$$IU(w) = \Sigma_j IU(BIT_j),$$

where BIT_j ranges over the objects of the BITs realized in w. Because BIT_j represents the object of the jth BIT, $IU(BIT_j)$ stands for IU(the object of BIT_j). To deal with some technical problems that may arise given uncertainty, the identity assumes standard names for worlds and objects of BITs. Section 3.2.3 explains this assumption and its purpose.[14]

The summation principle for a world's intrinsic utility is the result of a rule of coherence binding an ideal agent's degrees of intrinsic desire. A world's intrinsic utility depends on the world's relevant logical consequences. These are realizations of intrinsic desires and aversions. The intrinsic utility of a BIT's realization is the same in every world where the BIT is realized because its realization's intrinsic utility depends only on its realization's logical consequences, and they contribute the same amount of satisfaction in every world containing them. The intrinsic utility of the BIT's realization is therefore the right weight to give its realization in an assessment of the intrinsic utility of a world where it is realized. Also, in an assessment of a world's intrinsic utility, the BITs realized divide considerations into pros and cons without omission or double counting. Therefore, by the principle of pros and cons, the summation of intrinsic utilities of objects of BITs that a world realizes yields the world's intrinsic utility.[15]

---

[12] If, contrary to my practice, one also supposes that an intrinsic aversion to the negation of a proposition intrinsically desired is just as strong as the intrinsic desire that the proposition hold, then $IU(\sim Ha) = -IU(Ha)$, and $IU(\sim Wa) = -IU(Wa)$. Given this type of symmetry, the average intrinsic utility of worlds is zero.

[13] Feldman (1986: 218) advances a similar principle, one obtaining the intrinsic value of a world from the sum of the intrinsic values of the basic intrinsic value states it realizes.

[14] Summation of finite with infinite or infinitesimal intrinsic utilities requires nonstandard numerical analysis. So I assume standard, Archimedean intrinsic utility scales, where for every two intrinsic utilities the first is r times the second for some real number r.

[15] Sobel (1970: 403–5, 415) describes problems a summation principle such as mine faces. Use of BITs overcomes those problems. Also, Kagan (1988: 18–23) doubts

This sketch of the argument for the summation principle does not draw on Section 2.1.5's causal principles. I will elaborate it in a way that appeals to those principles, however. To introduce the structure of the more detailed argument, I present a familiar, analogous argument.

The weight of an object is a sum of the weight of the atoms it comprises. Its weight depends on the weight of molecules it comprises as well as on the weight of the atoms it comprises, but counting the atoms only does not omit the molecules. They are counted by counting the atoms. Also, counting atoms does not double-count any molecule. The tally of atoms must count twice all the atoms a molecule comprises for it to double-count the molecule. Because it counts each atom only once, it does not double-count any molecule.

The expanded argument for the summation principle has the same structure as this argument that an object's weight equals the sum of the weights of its atoms. It treats a world's intrinsic utility as a matter of the intrinsic utilities of the world's atomic and molecular components. It takes realizations of BITs as atomic and realizations of other ITs as molecular. The argument claims that by counting the BITs realized, one counts the realized causes of the ITs realized; and that by counting the realized causes of the ITs realized, one counts those ITs' realizations. Counting each BIT's realization exactly once prevents double-counting any IT's realization.

As the analogous argument assumes that the weight of a molecule equals the sum of the weights of its atoms, the argument for the summation principle assumes that the intrinsic utility of realizing a set of BITs equals the sum of the intrinsic utilities of the BITs' realizations. This assumption warrants the claim that counting atomic factors not only counts molecular factors but counts them in the right way. The assumption rests on the causal independence of BITs. Because a BIT is causally independent, it contributes the same amount of satisfaction in each of its realizations. Hence in the context of other BITs' realizations, it contributes a constant amount of satisfaction. The amount it contributes added to the amount they contribute yields the amount their combined realization contributes.

that there are moral reasons whose strengths are independent and additive. His doubts arise from examples of moral reasons that do not qualify. My account of BITs allays similar doubts about the existence of pros and cons whose strengths are independent and additive regarding $IU(w)$.

The argument for the summation principle begins by specifying the considerations relevant to the intrinsic utility of a world w. These are the considerations a summation of intrinsic utilities of BITs' objects must count exactly once. The relevant considerations for $IU(w)$ are considerations that influence the intrinsic attitude toward w, its direction and intensity. These are the objects of ITs realized in w that influence the intrinsic attitude toward w. Not every realized IT's object is influential. Imagine a world w in which p & $\sim q$ is true. Take an ID that $p \lor q$ arising from a BID that p and a BID that q. This ID has multiple realizations distinguished by the BIDs causing the ID that they realize. Only the ID's realization in w affects the intrinsic attitude toward w. Because its realization in w is the realization of the BID that p, p's realization matters but not $p \lor q$'s realization. Although the BID that q influences the ID that $p \lor q$, this factor behind the ID is irrelevant to the intrinsic attitude toward w. Its irrelevance makes $p \lor q$'s realization irrelevant.

According to CP3, an IT's realization in w also realizes some BITs causing it. The combined realization of those BITs is the IT's relevant realization. The IT toward the combined realization of those BITs' objects influences $IU(w)$. That IT's object is a relevant consideration. Every relevant consideration arises from an IT's realization in this way. I represent the combined realization of a set of BITs as a conjunction of the BITs' objects. If the set of BITs has a single member, the conjunction is degenerate and has a single conjunct. Under this representation, every relevant consideration is a conjunction of objects of BITs that w realizes.

To prevent omission of relevant considerations, the conjunction must include all the objects of realized BITs causing the ITs realized. It does not suffice to focus only on a set of BITs whose realization is sufficient for the IT's realization. Take the case of the two pleasures. Imagine BIDs for each pleasure and for constancy of pleasure, and consider the ID for the two pleasures' joint realization. Realizing the BID for each pleasure generates the pleasures' joint realization. Still, the BID for constancy affects the intensity of the ID for the pleasures' joint realization because their joint realization entails a constant level of pleasure. Focusing on the BIDs for the two pleasures thus misses a consideration relevant to the intrinsic utility of a world realizing the ID for their joint realization.

The argument for the summation principle shows that tallying BITs' objects neither omits nor double-counts any relevant consideration.

A crucial preliminary step is distinguishing counting from explicitly counting, and double-counting from explicitly double-counting. Although a tally of BITs' objects does not explicitly count conjunctions of BITs' objects (except degenerate conjunctions), perhaps it implicitly counts them. Also, although it does not explicitly double-count any conjunctions of BITs' objects, perhaps it implicitly double-counts some. The argument must eliminate these possibilities.

An object of an IT not explicitly counted twice in a tally of ITs' objects is double-counted anyway, and hence is implicitly double-counted, if and only if the IT's object is a logical consequence of each object of two ITs explicitly counted. For example, if there are IDs that p & q, that p & r, and that p, then counting both conjunctions doublecounts p.

An object of an IT not explicitly counted in a tally of ITs' objects is counted nonetheless, and hence is implicitly counted, if and only if the tally explicitly counts objects of ITs that together cause the IT. ITs that cause another IT fully give reasons for that IT (in a rational ideal agent). They account for the attitude's direction and intensity. So their objects may substitute for that IT's object in a tally of ITs' objects. For example, suppose an ID that p and an ID that q cause an ID that p & q. Then counting the ID that p and the ID that q counts the ID that p & q. Because counting is the opposite of omission, a tally of ITs' objects omits an IT's object if and only if it neither explicitly nor implicitly counts that object.

The summation for $IU(w)$ tallies objects of BITs that w realizes. The relevant considerations are conjunctions of BITs that w realizes. The summation does not omit or double-count relevant considerations. First, consider double-counting. Suppose the summation for $IU(w)$ double-counts some relevant consideration. According to the characterization of double-counting, the double-counted consideration is a logical consequence of each of two BITs' objects. Hence some conjunction of BITs' objects is a logical consequence of each of two BITs' objects. The conjunction, even if degenerate, is distinct from at least one of the two BITs' objects. That BIT's object entails the object of every BIT in the conjunction, at least one of which is distinct from it. However, according to Section 2.1.5's second causal principle, CP2, no BIT's object entails a distinct BIT's object. So double-counting does not arise.

Next, consider omission. Imagine a relevant consideration that is a nondegenerate conjunction of objects of BITs that w realizes and hence is not explicitly counted by a tally of objects of BITs that w realizes. The BITs whose objects are conjoined cause the IT toward the conjunction, according to Section 2.1.5's first causal principle, CP1. Hence tallying the objects of BITs that w realizes tallies the objects of the BITs causing the IT. Therefore, it implicitly counts the object of that IT, according to the characterization of implicit counting. Although the tally does not explicitly count that consideration, the tally still counts it.

Therefore, my method of calculating the intrinsic utility of a world neither omits nor double-counts any relevant consideration. All relevant realizations of intrinsic desires and aversions are counted by counting the BITs realized, and no relevant realization of an intrinsic desire or aversion is double-counted by counting the BITs realized. This justification of the summation principle by the principle of pros and cons is independent of the justification of expected utility analysis by that principle. It rests on an independent separation of considerations. As Chapter 3 shows, expected utility analysis does not break down the utility of a world as the summation principle breaks down the intrinsic utility of a world.[16]

2.3. COMPREHENSIVE UTILITIES OF WORLDS

Decision are guided by comprehensive utilities. Do intrinsic utilities yield comprehensive utilities? Yes, intrinsic utility is more fundamental than comprehensive utility. It can be used to compute comprehensive utility. In the simplest case, suppose that w is a name specifying a possible world, a maximal consistent proposition trimmed of irrelevant

---

[16] Lemos (1994) presents two counterexamples to a summation principle for intrinsic value. Are they also counterexamples to my summation principle for intrinsic utility? The first (p. 36), borrowed from Chisholm, imagines the state of affairs that Jones is pleased that Smith is suffering. To make my summation principle applicable, suppose that this state is the only relevant state of the world, and that Jones is intrinsically averse to it, has a BID for pleasure it realizes, and has no other realized BITs. Then my summation principle implies incorrectly that for Jones the world's intrinsic utility equals the intrinsic utility of realizing the BID for pleasure. This case makes an unwarranted assumption, however. Jones must have a basic intrinsic aversion that causes an intrinsic aversion to the bad pleasure. The world also realizes that BIT.

The second example (pp. 48–9, 63–4), borrowed from Parfit, concerns a world in which one lives a century of ecstasy and a world in which one lives a drab eternity. Allegedly, the first world may have higher intrinsic utility than the second,

detail. The proposition w entails everything relevant that would happen if it were realized. Because a world entails all its features, w's evaluation in terms of logical consequences is equivalent to its evaluation all things considered. Hence $U(w) = IU(w)$. The comprehensive utility of a world is its intrinsic utility. One can therefore obtain a world's comprehensive utility given a method of calculating its intrinsic utility. From the formula for a world's intrinsic utility, one may immediately derive an analysis of a world's comprehensive utility.

$$U(w) = \Sigma_j IU(\mathrm{BIT}_j),$$

where BIT_j ranges over the objects of the BITs realized in w. This is an intrinsic utility analysis of a comprehensive utility.

To illustrate the method of analysis, take Section 2.2.2's example about Adam and his intrinsic desires to be healthy and to be wise. If his intrinsic desire to be healthy is twice as strong as his intrinsic desire to be wise, then on a utility scale that takes the second desire as a unit $IU(Ha) = 2$ and $IU(Wa) = 1$. If the two desires are his only BITs, then the four realization patterns for them generate the relevant possible worlds. According to intrinsic utility analysis, $IU(Ha \ \& \ Wa) = IU(Ha) + IU(Wa) = 3$. The calculation for other worlds is similar. Their intrinsic utilities derive from the intrinsic utilities of the BITs they realize.

Chapter 5 extends intrinsic utility analysis to an option's comprehensive utility and explains how the analysis can accommodate uncertainty about the option's outcome. The intervening chapters introduce expected utility analysis as a general method of coping with uncertainty.

although the second has infinite intrinsic utility according to my summation principle's extension to the infinite case, assuming a constant amount of pleasure each day and a constant BID for it. I reject the putative counterexample's presumption that the intrinsic utility of the century of ecstasy is finite in comparison with the intrinsic utility of a day's pleasure in the drab eternity.

3

Expected Utility Analysis

To analyze a new safety standard's social utility, Section 1.1 divides that utility according to people the standard affects and according to goals those people have. When the new standard's outcome is uncertain, it is also helpful to divide its utility for a person according to its possible outcomes, perhaps, either a reduction or no reduction of workplace injuries. That utility is a weighted average of personal utilities of its possible outcomes, where the outcomes' probabilities supply the weights. This type of analysis is called *expected utility analysis*, because the weighted average is called an expected utility. It adds a new dimension to utility analysis, a dimension of possible outcomes.

This chapter presents a precise formulation of expected utility analysis. The new formulation incorporates several improvements over standard formulations. In particular, it defines an option's possible outcomes so that an agent's knowledge of their utilities requires only introspection. It also carefully attends to necessary idealizations, adopting and elaborating those Section 1.3.2 sketched. The next chapter defends and generalizes my version of expected utility analysis. Later chapters combine expected utility analysis with intrinsic utility analysis and group utility analysis.

My investigation of expected utility analysis assumes that expected utilities govern decisions. First, expected utilities generate options' utilities. Then the rule to maximize utility uses options' utilities to reach a decision. I therefore test my version of expected utility analysis by applying it to decision problems and verifying that it yields sound recommendations. Expected utility analysis governs conditional utilities as well as utilities. Because conditional utilities are connected with decisions less directly than utilities are, I use intuitions about conditional utilities themselves, not just decisions, to test the consequences of my version of expected utility analysis for conditional utilities.

Expected utility analysis asserts that an option's utility is a weighted average of the utilities of its possible outcomes. That is, *an option's utility is its expected utility*. As I interpret it, this formula expresses a principle of rationality, not a definition. By definition an option's expected utility is a probability-weighted average of the utilities of the option's possible outcomes. Its equality with the option's utility is a matter of rationality. Under the idealizations, rationality requires assigning utility so that an option's utility equals its expected utility. This section supports expected utility analysis. It is decision theory's best-known form of utility analysis.

Expected utility analysis breaks down the utility of an option in a decision problem according to the various possible outcomes of the option. For example, consider a gamble that pays $2 if heads comes up on a coin toss and $0 otherwise. The relevant possible outcomes are a gain of $2 and a gain of $0. Each possible outcome has a utility. The utility of a possible outcome is weighted by its probability. Then the products are added to obtain the expected utility of the gamble. The expected utility of the gamble is therefore 1/2 times the utility of gaining $2 plus 1/2 times the utility of gaining $0. Letting the utility of gaining $0 be 0, the expected utility of the gamble is 1/2 times the utility of gaining $2. (This need not equal the utility of gaining $1.) Expected utility analysis claims that an option's expected utility is its utility, the quantity that indicates the option's choiceworthiness according to the decision rule to maximize utility.

The possible outcomes of an option are usually divided by means of a partition of possible states of the world. For each state one uses the outcome of the option in that state. One may think of a possible outcome as the relevant aspects of the option's outcome given a possible state. The partition of states is chosen for convenience. Different partitions yield different sets of possible outcomes. In the example it is convenient to let the possible results of the coin toss yield the set of possible outcomes because then the outcomes are monetary gains.

Expected utility analysis separates considerations for and against an option in terms of subjective chances for possible outcomes of the option. Given the realization of an option, the agent has a chance for each of its possible outcomes. Each chance obtains with certainty, although each possible outcome has only a limited probability of obtaining; one is certain to have the chances for possible outcomes if one performs the option. Expected utility analysis is a way of dissecting an option's utility in terms of results that are certain to ensue. The division of chances according to possible outcomes generated by a partition of possible states guarantees that no chance is omitted or double-counted. Even if the relevant possible outcomes are the same for two members of the partition of states, the corresponding chances are different because the probabilities of the possible outcomes come from different states.

Consider a chance for an outcome. If the utility of the outcome is negative using indifference as a zero point, the chance is a con. If it is positive, the chance is a pro. The utility of a chance for a possible outcome is a fraction of the utility of the possible outcome itself. Intuitively, it is the product of the probability and utility of the possible outcome. Adding the utilities of the chances, or the products of the probabilities and utilities of the possible outcomes, is a way of adding pros and cons. An option's expected utility, the sum of the products of the probabilities and utilities of its possible outcomes, therefore yields the option's utility according to the principle of pros and cons. Thus the principle of pros and cons justifies expected utility analysis.

My application of the principle of pros and cons takes chances, not possible outcomes, as pros and cons. Also, the weights of the pros and cons are the utilities of these chances and not the utilities of the possible outcomes. Accordingly, the product of the probability and the utility of a possible outcome is not just an arithmetic pause on the way to the utility of the option. The product yields a pro's or con's weight, the utility of the chance for the possible outcome. Expected utility analysis divides an option's utility into the utilities of chances for the option's possible outcomes.

3.2. A PRECISE FORMULATION OF EXPECTED UTILITY ANALYSIS

Whereas the preceding section presents and justifies a rough formulation of expected utility analysis, a formulation that expresses the core of most versions of expected utility analysis, this section presents and defends a precise formulation. Greater precision, of course, leads to controversy. To separate issues, I first present a precise version of expected utility analysis under the restriction that states are independent of options. In the next chapter I remove this restriction and present a precise and completely general version of expected utility analysis. In assuming that states are independent of options, I assume both causal and probabilistic independence. This assumption is stronger than necessary, but postpones controversy about the best way to weaken it. I discuss the relevant issues in Section 4.2 when I remove the independence assumption to generalize expected utility analysis.

In presenting the restricted version of expected utility analysis, and later the general version, I assume the idealizations that Section 1.3.2 describes. More precisely, I assume that the decision maker is perfectly able to form utilities for options by expected utility analysis, and to maximize utility in forming preferences and making decisions. This requires the existence and accessibility of relevant probabilities and utilities, and certain cognitive gifts such as knowledge of mathematical truths and the ability to think instantly and without effort. I assume as well that when a decision maker applies expected utility analysis, he is in a fully rational state of mind so that the input for the analysis is rational and compensation for irrational input is not necessary. Finally, I assume that in his decision situation maximizing expected utility is feasible and that assumption of a decision does not alter expected utility comparisons. The precise content of these idealizations depends on the definition of expected utility adopted. Once a definition is adopted, however, the idealizations can be reformulated in terms of probabilities and utilities so that they are independent of the definition.

3.2.1. Options

An expected utility analysis breaks down the utility of an option into the utilities of chances for possible outcomes of the option. To specify a version of expected utility analysis, one must specify the chances whose utilities are added to obtain an option's utility, and the method of obtaining their utilities. I do this by considering the factors composing the chances tallied. Let me begin by discussing the nature of the options to which expected utility analysis applies. They are used to specify the relevant chances, and ultimately their expected utilities are compared to reach a decision.[1]

In decision problems it is standard to take options as actions, or possible actions, available to a decision maker. But it is advantageous to be more precise. Among an agent's possible actions are mental actions

---

[1] This section draws on Weirich (1983b).

78

such as calculating a sum or drawing an inference. A decision is also a possible mental action, the formation of an intention to do something. Decisions, although mental actions, are also physical. Decisions, along with other mental phenomena, have their place in a physicalist world view. According to some versions of physicalism, they are reducible to microphysical phenomena.

In a decision problem, it is best to apply expected utility analysis to the decisions that are possible resolutions of the problem rather than to the decisions' contents, the actions that might be chosen in resolving the problem. Taking options to be possible decisions, I say that an agent *adopts* or *realizes* an option. Saying that an agent *decides* on an option suggests a decision to make a decision, and hence a regress of decisions. But there is no regress. Agents may spontaneously decide. They may spontaneously realize a possible decision. Moreover, a spontaneous decision may be rational because deliberation may be unprofitable.

The main reason for taking options to be possible decisions is the temporal immediacy of a decision as opposed to an action chosen. A decision is the immediate resolution of a decision problem. An action chosen may not be performed until later. This means that a decision to perform an action may have consequences that the action lacks. For example, the decision may have mental costs that the action chosen does not have. One may not ignore such consequences of the decision.

My idealization about the cognitive capacities of the decision maker makes it unnecessary to worry about decision costs at the present stage of my theory's development. I assume that the decision maker has no cognitive limitations and hence that the cost of thinking and, in particular, deciding is zero. However, I want my theory's design to facilitate later removal of idealizations. After removing the idealization that thinking and deciding are costless, one can accommodate decision costs more easily if options are decisions. Then the consequences of an option include decision costs. Furthermore, even putting aside decision costs, a decision may have consequences not included among the consequences of the action chosen. For example, a decision to go shopping may cause one to ask a friend to come along. This consequence of the decision is not a consequence of going shopping. Also, a decision to jump off the high diving board might cause anxiety that the jump itself does not cause. An assessment of a decision's rationality must take account of such consequences of a decision. Taking options as decisions does the job neatly.

Because Section 3.2.3 takes an option's outcome comprehensively to include everything that would happen if the option were adopted, not just its consequences, accommodating a decision's consequences does not require taking options as decisions. The outcomes of the decision and the action decided on are the same, if we assume that one occurs if and only if the other does. But enlarging the framework for multidimensional utility analysis, to include noncomprehensive types of utility attuned to an option's consequences, requires addressing the issue. To make expected utility analysis extendible to causally restricted utility, it is best now to take options as decisions.

Another reason for taking options as decisions is that one is certain about one's ability to reach a decision. It is a mental action wholly within one's power, at least under my cognitive idealizations. On the other hand, one may be uncertain about one's ability to perform an action chosen, for example, going to Hawaii. Such an action requires the airlines' cooperation. When the ability to perform actions is uncertain, their choiceworthiness does not go by their expected utilities. The expected utility decision principle requires that uncertainty be restricted to states. Applying expected utility analysis to possible decisions, that is, certain mental actions, is a convenient way of restricting it to options that the decision maker is certain she can realize.[2]

I take an option as a possible decision, the formation of an intention to act. It is distinguished from other possible decisions by the content of the intention. The content of the intention is best given by a proposition. So it is convenient to take a possible decision to be represented by a proposition. Also, I attach utilities, or degrees of subjective desirability, to options and so to possible decisions. Because desirability attaches to propositions, it is again convenient to take a

---

[2] Given the idealizations, deliberation is instantaneous. So one need not worry about the possibility that an agent is uncertain that she can realize an option because she may, for instance, die before deliberation is over. Also, strictly speaking, the expected utility principle does not require that the agent be certain that she can realize each option. It suffices if she is certain for each option that she will realize it *if* she decides in an appropriate way. This conditional certainty is all the principle requires, and it can obtain even if deliberation and reaching a decision take time. An agent may be certain that if she decides in such and such a way she will realize such and such an option, even if she is not certain she will live long enough to be able to decide in that way.

To ensure that options are in an agent's direct control, Joyce (1999: 57–9) takes options, which he calls acts, as pure, present exercises of the will. These acts of will are similar to decisions, which are also in an agent's direct control.

possible decision to be represented by a proposition. I therefore take possible decisions, and so options, to be objects representable by propositions. For simplicity, I generally identify them with propositions. Specifically, I identify a decision that p with the proposition that the agent decides that p. For brevity, I sometimes represent that decision with the content proposition p, taking care to evaluate it as the decision with that content.

I take the set of all options to be the set of decisions that the agent can make at the time of decision. Because the word "can" is vague and ambiguous, the set of all options is also vague and ambiguous. To make that set more definite, I take the decisions that the agent can make at a time to be the propositions expressible by the set of declarative sentences of his language – the language of his thought, for example, the part of his mother tongue that he knows. This set is then limited by the agent's cognitive powers. So "can" has the sense of psychological possibility.[3]

A typical possible decision has as content a first-person action proposition with an upshot, such as that I go to the store. The first-person action sentence, "I go to the store," expresses this proposition. According to the theory of direct reference I adopt, the indexical "I" refers directly without the mediation of a Fregean sense. The agent to which it refers is a component of the proposition the sentence expresses.

Possible decisions with other types of content are generally ineffi-cacious. For instance, although someone could decide that $2 + 2 = 4$, or

---

[3] What counts as an option is still vague. It is something one *can* decide, but the rel-evant psychological ability is left vague. Of course, it depends on the agent's ability to form intentions to do such and such. This is influenced by his linguistic skills and other cognitive attributes. I do not attempt to provide details. I assume only that once the relevant sense of ability is made precise, the set of all options is well defined.

  The set of all options is generally a less inclusive set than the set of all proposi-tions because an option is a possible decision and thus meets psychological restric-tions. Even an agent ideal in my sense may think in a language that limits the possible objects of his decisions. Suppose, however, that some ideal agents may entertain and decide upon any proposition. Then the set of all options is the same as the set of all propositions. Is the set of all options still well defined? Set-theoretic paradoxes are sometimes used to argue against the existence of the set of all propo-sitions. Set theories providing for a universal set have principles that block the argu-ments, however. In any case, if pressed, I may dispense with the set of all options. All I really need are options that have maximum expected utility among all options. Whether all options form a set or not is immaterial.

decide to act so that $2 + 2 = 4$, nothing would follow from his decision. It is superfluous, however, to limit options to efficacious decisions. Inefficacious decisions have low utility and so generally give way to efficacious decisions in decision problems.

Also, I count as a possible decision in a degenerate sense the null decision – failure to reach a decision, or postponement of a decision. That is, I count as an option the decision with no content. This decision differs from the decision to do nothing. The latter has content, whereas the null decision does not. To obtain a propositional representation of the null decision, I assign an arbitrary necessary truth as its object. I identify the null decision with an inefficacious decision such as the decision that $p \lor \sim p$. The object of the decision to do nothing, in contrast, is the proposition that the agent is inactive.

The actions chosen may include conditional actions, or strategies. To take an example from Savage (1954: 13–15), a cook may decide to break a suspect egg into a saucer for inspection before either using the egg or throwing it away. Under my idealizations, a strategy selected never fails to be executed because of cognitive deficiencies, even though carrying out its stages requires decisions in addition to the initial decision to adopt it. One interesting feature of a decision to adopt a strategy is that it may have higher expected utility than a decision to adopt any of the unconditional actions it involves. For instance, the cook's strategy to break the suspect egg into a saucer and then act according to the results of its inspection has higher expected utility than both using the egg and discarding it – that is, actions unregulated by inspection of the egg. The expected utility of the saucer strategy is not just a probability-weighted average of the expected utility of the two actions it might produce, because those actions are tied to inspection results and have different expected utilities in the contexts that lead to them. They have higher expected utilities as possible outcomes of the strategy than they do as non-strategic actions.

My characterization of options bears on some issues concerning applications of expected utility analysis to options. Many traditional expected utility analyses treat an option as a lottery over a set of possible worlds where it is realized. If an option is a possible decision represented by a structured proposition, may an analysis of its utility treat it as such a lottery? A structured proposition cannot in general be identified with the set of possible worlds where it is true. Nonetheless, calculations of a structured proposition's utility may, if convenient, represent the proposition in terms of a set of possible worlds because propositions true in the same worlds have the same utility.

Next, consider expected utility analysis's application in decision problems. When the rule to maximize expected utility is applied to a decision problem, I assume that it is applied to the set of all options – that is, to all the decisions the agent might make in resolving the decision problem. This means that the rule to maximize expected utility is a lofty goal for decision makers, not easily attained. Humans are typically unable to entertain more than a few options. They cannot entertain, let alone analyze, all the decisions they might make. In fact, an important human decision skill is the ability to bring to mind unfamiliar but attractive options.

It might be tempting to make the expected utility principle more realistic by allowing it to apply to some subset of all the options available. But options meeting practical limits are often not suitable input for the principle. Take the set of options actually considered. The decision reached by maximizing expected utility with respect to it may be irrational. This may happen since a rational decision, as I take it, is comprehensively rational, not just rational given the options considered. A decision that maximizes expected utility among the options actually considered may fail to be comprehensively rational because the set of options considered is a set irrational to consider. The goal of maximizing expected utility may be generalized for cases where it is cognitively impossible to consider all options. Nonetheless, to simplify, I assume that agents can assess all options.

Dayton (1979) and many other theorists claim that the goal of rational decision is not to maximize expected utility with respect to all options, but rather to maximize expected utility with respect to an appropriate subset of options that are, among other things, mutually exclusive and jointly exhaustive. They believe that this restriction is necessary to guarantee that the goal does not discourage options entailed by options the goal recommends. They are afraid that without the restriction the goal might, for example, encourage buying a red hat because that action has maximum expected utility, but discourage buying a hat because that action does not have maximum expected utility. However, inconsistent advice can arise only if the goal is formulated so that it urges a decision maker to adopt an option *only if* its expected utility is maximal. I do not propose to formulate the goal that way. I prefer a formulation that simply urges a decision maker to adopt an option whose expected utility is maximal. Such a version of the goal does not discourage adopting an option of nonmaximum expected utility entailed by an option of maximum expected utility.

The weaker formulation of the goal is especially apt if options are taken as nonmental actions, as Dayton seems to take them. Then expected utility maximization is clearly not a necessary condition of rationality in ideal conditions. An action may be rational, although it does not maximize expected utility, if its realization is entailed by an action that does maximize expected utility.

Under my interpretation of options as possible decisions, the resolution of a decision problem comprises a unique option. Consequently, my formulation of the goal of rationality for decisions in effect urges a decision maker to adopt an option only if it maximizes expected utility. Its tolerance of nonmaximal options vanishes. Still, no inconsistency arises because the goal applies to possible decisions, and not also to all actions their execution entails. The goal does not recommend any decision while discouraging something the decision entails.

In any case, if options are taken as possible decisions, and possible decisions are understood so that an agent a realizes a possible decision that p at time t if and only if a decides that p at t, and p is all that a decides at t, then in a decision problem every option is incompatible with every other option. In particular, a decision to buy a red hat does not entail a decision to buy a hat. Furthermore, the members of the set of all options are mutually exclusive and jointly exhaustive and form an appropriate set of options for application of the expected utility principle according to Dayton's criteria. Consequently, Dayton's restriction does not narrow the set of options to which the expected utility principle applies.[4]

Given that options are possible decisions rather than actions that might be chosen, some questions arise concerning the relationship between rational decisions and rational actions. May it be rational to

---

[4] My formulation of the goal of rationality and my interpretation of options are just two ways of resolving Dayton's problem. Certain additional idealizations also resolve the problem. For instance, one might assume adoption of the best way of executing an option, so that the expected utility of buying a hat, taken as a lottery among the ways of realizing it, is the same as the expected utility of buying a red hat. A theoretical framework incorporating this idealization, however, cannot easily jettison the idealization for the sake of greater realism. Also, the idealization properly regulates the expected utilities of only the agent's actions, identified with their propositional representations, and not the expected utilities of propositions whose form of realization is out of his control. It does not ensure, for example, that maximum expected utility offers consistent advice about wishes.

make a decision but irrational to perform the action chosen? Conversely, may it be rational to perform an action but irrational to decide to perform it? A difference in the status of decision and action chosen is unusual but not impossible because the decision and action chosen are, after all, different objects of evaluation; their circumstances, including relevant alternatives, may differ. Consider the step from rational decision to rational action. Although thinking ahead helps a cognitively limited agent prepare for action, good habits for reopening deliberations prevent obstinacy. The reasons for adhering to one's decisions and for following one's plans may be overridden by changes in circumstances. Despite those reasons, acting according to one's prior decisions and plans may fail to maximize utility, and so may be irrational. Obviously, new information may explain why a decision rationally made should not be executed. A difference in status may also arise without a change in information. Kavka (1983) describes a case in which it is rational, if possible, to decide to drink a nonlethal toxin because of the reward the decision brings, but irrational to carry out that decision because the reward arrives before the decision's time of execution. Looking at the converse phenomenon, it may be rational to act spontaneously without a prior decision to so act. In a basketball game, a spontaneous drive toward the basket may succeed, whereas a prior decision to drive toward the basket may slow the drive and allow the defense time to react. The act is rational but a prior decision to perform it is irrational.

Another related issue is the relationship between reasons for deciding and for executing decisions. Normally, the reasons are similar. It is rational to decide upon an act because the decision generates the act (or at least makes it more likely). The reasons for the act provide the reasons for the decision. But suppose an agent who has good reasons to perform an act knows that because of akrasia or forgetfulness (reasons the idealizations put aside) he will not perform the act. Then a decision to perform the act is pointless and irrational. The reasons for the act do not transfer to the decision. But suppose further that a rich, eccentric psychologist will reward the decision itself. Then the decision has a point and is rational despite nonexecution. It is rational but for reasons that do not bear on the act.

These issues concerning decisions and acts are not problems for the view that options are possible decisions but rather are research topics that the view highlights. The literature cited in Chapter 1, note 10, pursues these issues. This section need not examine them further since the idealizations about agents put aside the difficult cases. Moreover, the nature of options, and the relationship of decisions and acts, are not pivotal matters for utility analysis. My characterization of options clarifies the decision rule to maximize utility, which I use to explain utility's role. My main topic, however, is utility analysis. Because the methods of utility analysis apply to acts besides decisions, their operation is independent of the nature of options in decision rules.

3.2.2. States

Next, consider the states that appear in expected utility analysis. Because subjective probabilities, or degrees of belief, attach to these states, I assume that they are, or can be represented by, propositions. Logical operations thus apply to states so that it makes sense to speak of a set of states that are mutually exclusive and jointly exhaustive. Expected utility analysis is to be applied with respect to a set of states that the agent is certain are mutually exclusive and jointly exhaustive. This ensures that no relevant consideration is omitted or double-counted. Normally, to obtain such a set of states, one uses a set of states that are logically exclusive and exhaustive, that is, a partition of states. When convenient, one may also use a set of states that are a priori exclusive and exhaustive. Either way, an ideal decision maker is certain that they are exclusive and exhaustive, because, as part of the idealization, the decision maker has full knowledge of all a priori truths, in particular, all logical truths.

In applications of expected utility analysis to decision problems, one need not use the same partition of states for each option. Expected utility analysis may apply to options one by one instead of all at once. A partition for an option may be selected so that the possible outcomes of the option with respect to the partition are as simple as possible to evaluate. Also, in applications of expected utility analysis propositions should be given standard names. The next section explains the reasons for this stipulation.

I contend, as do Jeffrey (1983: 78–81) and Joyce (1999: 121–2, 176–8), that expected utility analysis works with respect to any partition of states. There may be practical reasons for restricting expected utility analysis to certain partitions. Restrictions may make the utilities of possible outcomes easier to assess. But there is no theoretical reason for restricting expected utility analysis to certain partitions. The pros and cons argument for expected utility analysis justifies it for all partitions because every partition furnishes a means of separating considerations into pros and cons in a way that avoids omission and double-counting. To maintain the generality of expected utility analysis, I eschew restrictions on partitions.[5]

Because I sanction the application of expected utility analysis with respect to any partition, and because there are many partitions, there are many ways of computing an option's expected utility. I claim that an option's expected utility is its utility and recognize only one value for an option's utility. When there are multiple ways to compute expected utility, I claim that every way yields the same value. This follows from the pros and cons justification of expected utility analysis. As a consistency check, however, Section A.2 verifies that an option's expected utility is invariant with respect to the partition used in its computation. The proof obtains an option's expected utility with respect to a partition into worlds, and then demonstrates invariance by showing that for any partition the option's expected utility is the same as its expected utility with respect to worlds. The theorem shows the adequacy of the unrestricted form of expected utility analysis.

3.2.3. Outcomes

Possible outcomes are the third component of expected utility analysis. I characterize them with an eye to their role. Expected utility analysis attaches utilities to possible outcomes, utilities weighted and summed to obtain an option's utility. It advances a principle guiding the utility assignment of an ideal agent despite his uncertainty about the state of the world. The agent, if rational, complies knowingly. He is

---

[5] Some authors impose restrictions on the partitions used to compute expected utility. Lewis (1981) limits expected utility analysis to partitions of dependency hypotheses. Skyrms (1985) says that some versions of expected utility analysis are incoherent if applied to an option, given the option's negation as a state. Resnik (1987: 10) says that using the right partition is essential for an expected utility analysis, although he does not provide conditions for acceptable partitions. And Rawling (1993: sec. 7) considers whether causal decision theory differs from evidential decision theory in needing restrictions on partitions. Sobel (1994: chap. 9) provides a thorough discussion of the issue. He shows that restrictions on partitions are necessary for some versions of expected utility analysis, but unnecessary for others; in particular, they are unnecessary for the version of expected utility analysis presented here, in his terminology *Partitions-for-U*. Section 4.2.1 addresses, in particular, Skyrms's worry about using negations of options as states. It shows that in my version of expected utility analysis such states do not generate incoherence.

aware of the utilities he assigns to options and possible outcomes and makes them comply. I therefore characterize outcomes so that their utilities are accessible to an ideal agent despite uncertainty about the state of the world. Given an application of the expected utility principle, in which an option and its possible outcomes are named, the agent recognizes the options and outcomes as named and makes his utility assignment conform with the principle.[6]

To introduce the outcomes that arise in expected utility analysis, I start with some points about an option's outcome and then extend them to an option's outcome, given a state. Those conditional outcomes are featured in an expected utility analysis. Although I do not always say so explicitly, the conditional outcomes are possible outcomes because most options and states are not realized.

Rational deliberation is forward-looking. It considers how possible decisions affect the future. To focus attention on an option's effect, it is common to interpret an option's outcome as all the option's causal consequences. Then events an option does not affect are not evaluated. Deliberations may neglect their utility because they provide a common background for all options and so do not influence an option's standing with respect to other options.

The causal interpretation of an outcome is economical but creates analytical problems. It requires separating the events that would occur if an option were realized into events that are consequences of the option and those that are not. This is a difficult task. Many well-known problems plague attribution of causal responsibility – for example, problems of overdetermination. Although we have rough-and-ready intuitions about causal responsibility, a complete theory of causal responsibility is a tall order. To table issues concerning causal responsibility, this chapter adopts a noncausal interpretation of outcomes for its version of expected utility analysis. Once metaphysical issues about causal responsibility have been resolved, a compatible causal form of expected utility analysis can be articulated.

One attractive noncausal account of outcomes takes an option's outcome to be the option itself together with all simultaneous and subsequent events. This is a temporal segment of a possible world that I call the *aftermath* of the option. The evaluation of aftermaths may consider the context the past creates. Sometimes that context affects an

---

[6] See Vallentyne (1987: 57–60; 1988: 91–4) for a survey of interpretations of outcomes taken as objects of evaluation in teleological theories of morality.

aftermath's utility. For instance, if my family has farmed a particular plot of land for generations, then the utility of my farming it may be greater than it would be otherwise. When I evaluate my act's aftermath, I take account of the past. But once I appraise each option's aftermath, I may compare options by aftermaths. I do not have to evaluate their common past also. Because events prior to a decision are the same for every option, their evaluation does not influence the ranking of options.

Unfortunately, taking outcomes as aftermaths also raises some complex analytical issues. First, to identify a proposition's aftermath, one must identify its time of realization. This problem is difficult for some propositions, and so plagues a general formulation of expected utility analysis. It is, however, tractable for propositions expressing options. Because I take options as possible decisions, the temporal location of any option's realization is the moment of decision. Still, a second problem remains. To evaluate an option's aftermath, one must identify the events realized in it, in particular, the basic intrinsic attitudes realized in it. Because, for a complex event realized over a period of time, it may be hard to identify a moment when it is realized, it may be hard to determine whether it is part of an option's aftermath. Fixing a temporal location is difficult for some actions – for example, becoming a philosopher. Does it happen when one declares a major in philosophy, when one enters a graduate program in philosophy, when one accepts a post in a philosophy department? It's hard to say. Also, propositions expressing other events of interest seem to lack definite times of realization. Take, for example, generalizations about the future, such as that one lives forever or that everlasting peace is achieved. If realized, when are they realized? Or, suppose that an agent wants to have many friends and succeeds. Because "many" and "friend" are not precise, the object of the desire is a vague proposition. It is hard to pick out a precise interval to which to assign satisfaction of the desire. The agent's whole life is big enough, but too big.

To make its proposals independent of such complications, this chapter interprets outcomes independently of temporal distinctions as well as independently of causal distinctions. Its interpretation of an option's outcome does not require identifying the option's consequences or its aftermath. It takes an option's outcome comprehensively so that the outcome includes everything that would occur if the option were realized, that is, the option's past, present, and future – the option's world – not just its causal consequences or its aftermath. For example, the outcome of having fire insurance if there is a fire includes the fire itself as well as the net compensation the insurance provides for fire damage. It includes the consequences of having the insurance if there is a fire, but also everything else that would be true if one had fire insurance and there is a fire. It includes even events prior to the purchase of the insurance, events going back to the beginning of the world and clearly not caused by having fire insurance.[7]

Let me clarify the reasons for not using temporal and causal distinctions to reduce the considerations pertinent to an option's assessment. As a contextualist, I have no objection to defining outcomes using temporal or causal distinctions despite problems operationalizing those distinctions. In fact, this chapter's and the next's treatment of expected utility analysis does not forgo all use of temporal and causal distinctions. Causality, for instance, has a role in Section 4.2's general formulation of expected utility analysis. It arises in characterizations of appropriate probabilities for option-state pairs. Moreover, reliance on comprehensive utility does not really dispense with causality in utility analysis. It merely buries causality in the outcomes of propositions evaluated instead of displaying it openly. To calculate the comprehensive utility of reporting a criminal, say, the agent must consider whether, if he reports, he will be in a world where he has causal responsibility for apprehension of the criminal and so is entitled to a reward. Features of the option's outcome raise causal issues, even if the type of utility applied to the option is not causally defined. The gain from

---

[7] To simplify treatment of outcomes, this section assumes that an option's world exists. More generally, for every proposition p, it assumes that p's world exists. It is the nearest p-world. If p is counterfactual, there may be no such nearest p-world. Ties for nearest world may arise, for example. But this section ignores such complications. Section 4.2.1 takes some steps to handle them.

This section's assumption is not a large simplification in the context of expected utility analysis. Many cases meet the assumption. Because worlds are trimmed of irrelevant detail, ties among them are rare. Furthermore, expected utility analysis focuses on a option's world given a state, and states are selected to break ties. For example, consider the utility of betting a sum of money on rolling a six with a die. If one declines the bet and does not roll the die, no world is the bet's world. It is indeterminate whether one would have won or lost the bet. However, given the state that one rolls a six, the bet's world is determinate. It is a world in which one wins the bet. Cases that underdetermine an option's world may still determine an option's world, given a state. Finally, recall the idealization that the probabilities and utilities an option's expected utility analysis requires exist. It includes the assumption that an option's supposition is at least sufficiently determinate to generate those probabilities and utilities. So that idealization reduces indeterminacy concerning an option's world.

90

avoiding causal utility is not conceptual economy in the assessment of utility, but merely conceptual economy in the definition of utility. Reliance on comprehensive utility does not yield a net conceptual economy for decision theory. This chapter dispenses with temporal and causal distinctions when characterizing outcomes only to skirt ancillary metaphysical issues about events.

Taking outcomes comprehensively makes expected utility analysis assess the past and other events causally independent of options. These considerations seem irrelevant to decision making, but they are harmless. They do not disrupt expected utility analysis given a suitable companion interpretation of an option's utility. In accordance with standard practice, I take $U(o)$ as a comprehensive assessment of o. Expected utility analysis using comprehensive outcomes then generates an option's utility. For instance, when s is a logical truth, and constitutes a partition of states by itself, $U(o) = U(o$'s outcome given $s)$ by expected utility analysis, and the second utility equals $U(o$'s outcome). The implied identity $U(o) = U(o$'s outcome) holds given my idealizations when o's outcome is taken comprehensively, because under my interpretation U evaluates propositions comprehensively.

Even if one did not adopt a comprehensive interpretation of an option's utility, deciding by maximizing expected utility, defined in terms of comprehensive outcomes, would still be sensible. Comprehensive outcomes would carry excess baggage but would not change the decision procedure's final destination. Although an expected utility computed using an option's possible comprehensive outcomes would not yield the option's utility taken noncomprehensively, it would still yield an accurate measure of the option's choiceworthiness. Taking possible outcomes comprehensively introduces considerations besides the determinants of $U(o)$ narrowly construed. But those extra considerations have the same effect on the expected utility of all options in a decision problem and so no effect on the expected utility ranking of the options.

My version of multidimensional utility analysis may be expanded to make this explicit. To it, one may add temporal and causal dimensions of analysis. It may then incorporate temporal and causal variants of utility once the metaphysical issues on which they depend are resolved. Switching between types of utility amounts only to changing the scale for assessments of utility. To move from an option's comprehensive utility to its causal utility, one puts aside basic intrinsic attitudes (BITs) whose realization the option does not influence. To move from an option's comprehensive utility to its temporal utility, one puts aside BITs whose realization occurs prior to the option's time. Those moves amount to subtracting a constant from the option's comprehensive utility, the same constant for all options, because options, as I take them, are possible decisions with the same time and place of realization. Such scale changes do not affect the options' ranking. They just simplify the procedure for ranking options by reducing the scope of the considerations on which the ranking rests. They remove considerations common to all the options in a decision problem.[8]

I have adopted an interpretation of outcomes that makes them comprehensive. They comprehend the past and other unavoidable events. What type of entity are outcomes? Expected utility analysis attaches utilities to outcomes. Because utilities attach to propositions, I take outcomes to be, or to be represented by, propositions. This distinguishes outcomes from concrete events. A concrete event may be expressed by many propositions, not necessarily having the same utility.

It is common to take an option's outcome to be the world that would be realized if the option were realized, that is, the option's world. That interpretation of outcomes is the distinctive feature of a decision theory Sobel (1994, chap. 1) calls world Bayesianism. This tradition suggests that I take an option's outcome as a possible world trimmed of irrelevant detail, a maximal consistent proposition simplified to address only relevant matters. My interpretation of outcomes departs from this tradition, however, in response to recent research on the propositional attitudes.

States and outcomes are the objects of probability and utility assignments. Because probability and utility assignments rest on belief and desire, they involve propositional attitudes. The literature on propositional attitudes suggests that taking the objects of probabilities and utilities to be propositions oversimplifies the underlying attitudes. As I review the argument, I paint in broad strokes, although theories of the propositional attitudes are subtle and complex. Because my principal topic is utility analysis, I address only essential points.

Frege's puzzle about belief begins the argument. Because Cicero is Tully, how is it possible to believe that Tully is Tully without believing

---

[8] Gibbard and Harper (1981: 167) briefly present the justification for ranking options by causal utilities. They claim that the utility of an option is the sum of the utility of the consequences of the option and the utility of the unavoidable results of the option.

that Cicero is Tully? Frege's answer is that the object of a belief is a proposition, and a proposition is structured, having senses as components. The proposition that Tully is Tully differs from the proposition that Cicero is Tully. The theory of direct reference upsets this response to the puzzle. According to that theory, proper names refer directly and not via a sense. A sentence with a proper name expresses a structured proposition that has the name's denotation as a component. The proposition that Tully is Tully has Cicero as a component. It is the same as the proposition that Cicero is Tully. If we grant this point, the objects of propositional attitudes must be more fine-grained than propositions.

Crimmins (1992) purchases finer grain by attending to the context of belief reports. Suppose one reports that Caius believes that Tully is Tully. The report makes tacit reference to a mental representation by which Caius grasps the proposition Tully is Tully. It implies that he believes the proposition via a thought with the structure of the sentence that the report uses to express the proposition. The report may be true whereas the report that Caius believes that Cicero is Tully may be false. The two reports tacitly refer to different ways of grasping the same proposition. Caius may grasp the proposition one way but not the other way.[9]

According to Crimmins, belief is a relation between an agent, time, proposition, and mental representation. A belief report is true if the content proposition is believed by means of the particular mental representation to which the report tacitly refers. Probability and utility assignments treat belief and desire as relations between an agent, time, and proposition. They ignore the mental representations by which the agent grasps propositions. This simplification makes probability and utility theory elegant, but it needs justification in light of the fine grain of attitudes.

Strictly speaking, reports of propositions' probabilities and utilities tacitly refer to ways of grasping propositions. The way in which a report names a proposition indicates the relevant mental representation. For instance, the report "Caius assigns probability 1 to the proposition that Tully is Tully" tacitly refers to a mental representation indicated by the that-clause naming the proposition that Tully is Tully. The report may be true because Caius assigns maximum probability to the proposition under that mental representation. On the other hand, it may be false

---

[9] Weirich (forthcoming: sec. 2) provides a more detailed but still brief account of Crimmins's view. Richard (1990) presents a view similar to Crimmins's.

to report "Caius assigns probability 1 to the proposition that Cicero is Tully." Caius may not assign maximum probability to the proposition under the mental representation that the second report indicates. Caius may not know that the sentence "Cicero is Tully" names the same proposition as the sentence "Tully is Tully."

To make reports of probabilities and utilities independent of the ways in which the reports name propositions, I require that they use standard names of propositions. A standard name for a proposition is a sentential name, one that fully specifies a proposition and makes its denotation transparent to an ideal agent (despite ignorance of some a posteriori matters). Because "the proposition that Cicero is Tully" is not a standard name for the proposition that Tully is Tully, I ignore the report that for Caius the probability that Cicero is Tully is less than 1. Given the restriction to standard names, reports of probabilities and utilities agree. For instance, $\ulcorner U(p) = U(q) \urcorner$ is true if $\ulcorner p \urcorner$ and $\ulcorner q \urcorner$ are standard names of the same proposition. The restriction to standard names makes the influence of the names of a proposition uniform and so an unrepresented background feature. Under the restriction, one may treat a proposition as having a probability and utility independent of the way it is named.[10]

Also, in probability and utility laws, strictly speaking, propositional variables are placeholders for names of propositions and are not directly referential. The fine grain of attitudes requires that instances involve names indicating mental representations. To simplify, I specify that instances involve standard names of propositions. Given this policy, the laws function as if the variables were directly referential. Appealing to standard names simplifies utility analysis so that it may treat probability and utility as one-place functions of propositions rather than as two-place functions of propositions and mental representations.[11]

---

[10] A name for a proposition, such as a name with an indexical, may be standard for one agent but not for another agent. Perhaps for some agent, a proposition has no standard name. Then I do not apply expected utility analysis to the agent and proposition. Also, for simplicity, I assume that for an ideal agent a standard name for a proposition indicates an appropriate mental representation of the proposition in as much detail as matters for any report of the proposition's probability and utility.

[11] Horgan (2000) relies on standard, or canonical, names for his account of the probabilities at work in decision theory. He takes standard probabilities to attach to statements involving canonical names. Nonstandard probabilities attaching to statements involving noncanonical names create paradoxes, he says. For another account of standard names, see Kaplan (1969).

According to the simplification policy, a proposition's probability and utility are relative to standard names for the proposition. Given a standard name for the proposition, an ideal agent knows its probability and utility. But given a nonstandard name for the proposition, the agent may not know its probability and utility.

To see how this ignorance may arise, consider an analogous case: height as a function applied to people. There are three ways one might be ignorant of the denotation of $\ulcorner H(p) \urcorner$. One might be ignorant of the height function, at least in the case where its argument is p. One might know the height function but not know the denotation of $\ulcorner p \urcorner$. Or one might know the height of all people by their given names, know that $\ulcorner p \urcorner$ denotes the shortest spy, and yet not know the shortest spy's given name. An ideal agent knows his utility function applied to propositions standardly named but may not know which proposition standardly named is the proposition nonstandardly named by $\ulcorner p \urcorner$, and so may not know the denotation of $\ulcorner U(p) \urcorner$. A proposition has one utility (assigned with respect to standard names of the proposition) but presents itself in various guises. An agent may not recognize a proposition in one of its guises and so may not know its probability or utility under that guise.

Because a rational ideal agent may not recognize a proposition given a nonstandard name for it, she may violate probability and utility laws in instances generated by replacing propositional variables with nonstandard names. For example, suppose a law generates an instance $\ulcorner U(p) = x \urcorner$ in which $\ulcorner p \urcorner$ is not a standard name for a proposition. The equality may not hold because the agent does not assign the proposition p the utility x under the mental representation $\ulcorner p \urcorner$ indicates. Ensuring compliance with probability and utility laws requires replacement of variables with standard names of propositions so that the agent knows the probabilities and utilities of the propositions named. The probabilities and utilities in the laws' instances must be known for the laws to bind.

Reliance on standard names for propositions is my method of attaching probabilities and utilities to propositions, making probabilities and utilities accessible to an ideal agent, and ensuring a rational ideal agent's compliance with probability and utility laws. So specifying an option's outcome requires specifying a standard name for a proposition. The right name has to be an expression of everything that would hold if the option were realized but must also be selected to ensure a rational, ideal agent's knowing compliance with expected utility analysis. The name (in the agents' language) must be constructed from a name of the option, or a name of the option and a state of the world, according to a routine procedure so that applications of expected utility analysis are routine and not subject to uncertainty. Moreover, the name must be such that an ideal agent knows the proposition named despite uncertainty about the state of the world. This ensures that, given the proposition's name, he knows the proposition's utility.

The goal of comprehensiveness suggests taking the outcome of an option o as a possible world trimmed of irrelevant detail, a maximal consistent proposition simplified to address only relevant matters. To obtain a name for the world from the name for the option, call it the world that would be realized if o were realized, o's world for short. Unfortunately, an agent is generally uncertain of an option's outcome so conceived. He is uncertain of the trimmed world that would obtain if he were to realize the option. So he is uncertain of the utility of the option's outcome taken as the trimmed world named according to the formula. But under my idealizations he should know the utility of o's outcome. He should know $U(o)$, and, according to my interpretation of $U(o)$, it is the same as the utility of o's outcome. So he should know $U(o$'s outcome$)$. I need an interpretation of o's outcome that allows for knowledge of $U(o$'s outcome$)$ by allowing for knowledge of the proposition that is o's outcome as named according to a formula for expressing it.

The problem is that $\ulcorner o$'s world$\urcorner$ is not a standard name for a proposition. It is not a sentential name specifying a maximal consistent proposition. Thus an ideal agent may not know which maximal consistent proposition it names. Consequently, he may not know his utility assignment for o's world so named. According to my policy for preventing such ignorance, I must specify a standard name for an option's outcome, a name that transparently denotes the proposition constituting the outcome. I use the name $\ulcorner$the proposition that o's world obtains$\urcorner$. This name is sentential and so standard even if o's world is unknown. The proposition it names is nonmaximal but specified. It serves as o's outcome.

To bring out the difference between a proposition's outcome and its world, I introduce some notation. Let $\ulcorner W(p)\urcorner$ stand for p's world, the possible world that would obtain if p were realized, and let $\ulcorner O(p)\urcorner$ stand for p's outcome, the nonmaximal proposition that $W(p)$ obtains. $\ulcorner W(p)\urcorner$ is a nonstandard name for a proposition, whereas $\ulcorner O(p)\urcorner$ is a standard name for a proposition, and so a specification of a proposition. In a typical case where the expression $\ulcorner p\urcorner$ names a nonmaximal proposition, the expression $\ulcorner W(p)\urcorner$ names but does not specify a possible world. The expression's denotation is unknown. $U(W[p])$ is therefore also unknown, even if the utility of every possible world is known. $U(W[p])$ is an unknown quantity typically constant with respect to changes in information about the denotation of $\ulcorner W[p]\urcorner$, because such information typically does not affect the utility of worlds. On the other hand, the expression $\ulcorner O(p)\urcorner$ specifies a nonmaximal proposition. Its denotation is known and has a known utility. $U(O[p])$ is an estimate of $U(W[p])$. It is an expected utility, a probability-weighted average of the utilities of the worlds where p holds, the worlds that might be named by $\ulcorner W(p)\urcorner$. Thus $U(O[p])$ may vary with information bearing on the denotation of $\ulcorner W(p)\urcorner$.

To make the utility of p's outcome accessible as well as comprehensive, I take p's outcome to be named by $\ulcorner O(p)\urcorner$ rather than by $\ulcorner W(p)\urcorner$. The name for p's outcome is obtained by applying a routine formula to p's name. An ideal agent knows the proposition $\ulcorner O(p)\urcorner$ expresses and hence knows $U(O[p])$ despite uncertainty about the denotation of $\ulcorner W(p)\urcorner$. Substitute names of p's outcome are not authorized for use in applications of utility laws unless they are also specifications of $O(p)$.

Consider how my way of interpreting and naming outcomes works in utility laws. I take utility laws as schemata generating applications in which the laws' placeholder variables are replaced with names. Along with a law I assume an application routine, that is, a routine for obtaining names to replace variables. A utility law expresses a constraint to which a rational ideal agent knowingly conforms. Her conformity arises from knowledge of the denotation of names for propositions used in applications, so the names used are significant. Take the simple law that $U(p) = U(O[p])$. I have this law in mind when, to express the comprehensiveness of utility, I say that the utility of a proposition is the utility of the proposition's outcome. The law is a schema for applications rather than a generalization involving directly referential variables. An application follows the routine specified for obtaining the name for $O(p)$ from the name for p. Call that routine O. A fuller expression of the law is, "For a rational ideal agent, knowingly, $U(p) = U(O[p])$, where 'p' is replaced with a standard name for a proposition and '$O(p)$' is replaced by the standard propositional name generated from p's name by the routine O."

The context for a proposition's name in applications of utility laws is doubly intensional, involving utilities and knowledge of utilities. Within the epistemic context created by a utility law's application, the way a proposition is named matters. Standard names produce accessible utilities, which are required to ensure knowing compliance with the law. For an ideal agent, a standard name of a proposition transparently denotes the proposition. In the example the variable "p" is replaced not with just any name of a proposition, but with a name that specifies a proposition, a standard name. The expression "$O(p)$" is also replaced with a standard name. Replacing "p" or "$O(p)$" with a nonstandard name may not yield a true instance of the law. The propositions named by $\ulcorner p \urcorner$ and $\ulcorner O(p) \urcorner$ have the same truth value even if $\ulcorner p \urcorner$ is a nonstandard name. But both names must be standard to guarantee that the propositions named are logically equivalent and so are rationally constrained to have the same utility. The name $\ulcorner O(p) \urcorner$ is invariably a standard name for a proposition. It is a standard name even if the name replacing "p" is not standard. However, to make standard all propositional names involved in applications of the law, I stipulate that "p" be replaced with a standard name.

For contrast, notice that it is not a utility law that $U(p) = U(W[p])$. In applications the proposition that $\ulcorner W(p) \urcorner$ names is maximal and not logically equivalent to p if p is nonmaximal. Also, the naming routine for $W(p)$ does not generate a specification of a proposition. Hence $U(W[p])$ may be unknown even when p is standardly named and $U(p)$ is known. Its unknown value does not constrain $U(p)$'s value, so these values need not be identical. If a standard name replaces "$W(p)$" as well as "p," identical utilities are still not ensured because $U(p)$ is a probability-weighted average of worlds and need not equal the utility of a single world.

Now consider outcomes in expected utility analysis. Expected utility analysis uses conditional utilities. Each is the utility of an option given a state. The conditional utilities are formulated with an eye toward extensions of expected utility analysis to nonideal cases. However, given the idealizations the conditional utilities equal utilities of conditional outcomes. Each is the utility of the outcome of an option given a state. The utilities of these conditional outcomes are themselves nonconditional. My justification of expected utility analysis under the idealizations focuses on the utilities of the conditional outcomes. Because the principle of pros and cons takes utilities of considerations with respect to the same information, the analysis must use nonconditional utilities of conditional outcomes. This section explicates the conditional outcomes. Section 4.2.2 explicates the conditional utilities.[12]

First, I extend the interpretation of outcomes to conditional outcomes. The main issue is knowledge of conditional outcomes. To work, expected utility analysis must confine uncertainty to states. It must curtail uncertainty concerning outcomes given states. The characterization of conditional outcomes must ensure that the decision maker is certain of the outcome of an option given a state. I take the outcome of an option given a state to be *the proposition that the possible world obtains that would obtain if the state holds and the option were realized.* The italicized name is a standard name for the proposition it denotes. "The possible world that obtains if the state holds and the option were realized" may have an unknown denotation. So the utility of the proposition so named may have an unknown value. But "the proposition that the possible world obtains that would obtain if the state holds and the option were realized" has a known denotation, and the utility of the proposition so named has a known value.

To extend earlier notation to conditional outcomes, I use $\ulcorner W(o$ given $s)\urcorner$ to stand for the possible world that obtains if s holds and o were realized. Then I use $\ulcorner O(o$ given $s)\urcorner$ to stand for the proposition that $W(o$ given $s)$ obtains. In utility calculations, a single possible world represents $W(o$ given $s)$, whereas a set of possible worlds represents $O(o$ given $s)$, the set of possible worlds that might be $W(o$ given $s)$. The proposition $W(o$ given $s)$ is maximal. Typically, the denotation of the expression $\ulcorner W(o$ given $s)\urcorner$ is unknown. The proposition $O(o$ given $s)$ is nonmaximal. The routine for naming it generates a specification of the proposition. The denotation of the expression $\ulcorner O(o$ given $s)\urcorner$ is known. An ideal agent may not know the world signified by $\ulcorner W(o$ given $s)\urcorner$ but he knows the proposition signified by $\ulcorner O(o$ given $s)\urcorner$. Hence $U(O[o$ given $s])$ is accessible to him. The utility he attaches to $O(o$ given $s)$ is his estimate of the utility of $W(o$ given $s)$. His estimate may vary with changes in information without any changes in goals. $U(O[o$ given $s])$ equals a probability-weighted average of the worlds that might obtain if o were performed given that s obtains. As information changes, the probability weights change, and the value of $U(O[o$ given $s])$ changes as well.

---

[12] See Weirich (1982) for the advantages of using $U(o$ given $s)$ rather than $U(O[o$ given $s])$ in expected utility analyses in nonideal circumstances where desires are uncertain, and $U(o$ given $s)$ does not equal $U(O[o$ given $s])$.

I have characterized utility comprehensively so that under my idealizations the utility of an option given a state is the utility of the outcome of the option given the state. That is, $U(o$ given $s) = U(O[o$ given $s])$, provided that o and s are standardly named and the routine for naming $O(o$ given $s)$ is followed. I assume standard names in the context of utility laws but often do not state this assumption explicitly. Relying on the assumption, I generally conflate the distinction between propositions and their standard names. Having shown that appeal to standard names permits treating probabilities and utilities as functions of propositions, I make this simplification.

To summarize, my interpretation of outcomes takes them as comprehensive rather than causally or temporally limited and, although it takes them as comprehensive, allows them to be nonmaximal propositions. In expected utility analysis, it represents them with sets of possible worlds, but not necessarily single possible worlds. The main reason for not taking an option's outcome as a world is an agent's ignorance of a posteriori matters, including the world the option realizes.

3.2.4. Probability and Utility

My presentation of expected utility analysis now turns to probability and utility. They are both taken as psychological quantities, that is, as subjective or personal rather than objective quantities. They depend on an agent's beliefs and desires, not other facts about the world, such as relative frequencies and intrinsic values. Two rational agents may assign probabilities and utilities differently. So probability and utility assignments are relative to agents. Although rationality imposes some constraints, it leaves agents room to exercise their tastes in adopting goals and assessing the force of evidence. One agent may like swimming and another may dislike it. Their utility assignments for swimming may then differ. One agent may find a witness convincing whereas another finds the witness unconvincing. Their probability assignments for the witness's claims may then differ. Subjective probabilities depend on beliefs. Subjective utilities depend on desires and beliefs. The subjective utility of a lottery ticket, for example, depends not only on the desirability of the prize but also the prospect that the ticket will win.

Subjective interpretations of probability and utility suit decision theory better than objective interpretations. Probability taken objectively and as physical, for instance, makes decision principles hard to apply because physical probabilities – say, relative frequencies – are often unknown. A decision maker needs a kind of probability that is readily available for use in solving decision problems. Similar points about accessibility motivate the adoption of subjective utility for decision principles.

I take probability and utility to be rational degree of belief and rational degree of desire respectively. This makes probability inductive or epistemic because rational degree of belief depends on evidence, and it also makes utility epistemic because rational desire depends on evidence as well as personal ends. Despite being subjective quantities, probabilities and utilities obey certain laws, because rationality requires that degrees of belief and desire meet certain constraints. Probabilities, for instance, obey the standard laws of the probability calculus.

There is one respect in which treating rational degrees of belief as probabilities may be misleading. Standard formulations of probability theory include certain existence laws for probabilities. For example, it is assumed that if both p and q have probabilities, then p & q has a probability. These existence laws are not principles of rationality. So they are not necessarily satisfied by rational degrees of belief, even given my idealizations. When I say that rational degrees of belief satisfy the basic laws of probability, I exclude the existence laws. I mean only these versions of Kolmogorov's axioms: (a) $P(s) \geq 0$, (b) $P(s) = 1$ if s is an a priori truth, and (c) $P(s \vee s') = P(s) + P(s')$ if s and s' are mutually exclusive a priori. They govern only probabilities that exist. For instance, if s and s' are mutually exclusive a priori, and $P(s \vee s')$, $P(s)$, $P(s')$ all exist, then the first probability equals the sum of the other two.

I take degree of belief and degree of desire as primitive, unlike the operationist. But taking them as primitive means only taking them as undefined in the strict philosophical sense; it does not mean presenting them without an introduction. My brand of empiricism, contextualism, calls for careful introduction of primitive concepts. I clarify degree of belief and degree of desire by placing them in a context. Subjective probability and utility have held roles in theories of practical reasoning for hundreds of years. These theories elucidate them. I assume acquaintance with subjective probability and utility from their histories. So the introduction of degree of belief and degree of desire provided here is abbreviated. I just point out features of them that are important for my theory and are distinctive of my particular interpretation of them.

One feature to note immediately is that utility as I interpret it differs from utility as it appears in classical utilitarianism. According to my interpretation, utility is not pleasure or happiness, or the balance of pleasure over pain, or of happiness over unhappiness. It registers all desires and aversions, even altruistic desires for another person's happiness. Decision theory demands that utility have this broad interpretation.[13]

Before introducing degree of belief and desire further, let me address a worry some may have. Some may ask how quantities can be primitive. Quantities seem to require definition. This impression may have two sources. First, one may think that quantities cannot be observed and so must be defined in terms of something that is observable. This objection rests on the operationist theory of meaning rejected in the first chapter. It has no force. Second, one may think that quantities require conventions to establish scales for them and so cannot be primitive. My introduction of degrees of belief and desire presumes the standard scales for degrees of belief and desire: a ratio scale for degrees of belief ranging from 0 to 1, and an interval scale for degrees of desire. In taking degrees of belief and desire as primitive, I do not assume that the scales for them are primitive, only that ratio comparisons of degrees of belief and interval comparisons of degrees of desire are primitive. These ratio and interval comparisons are not conventional and so may be taken as primitive.[14]

To begin filling out this account of probability and utility, I take degrees of belief and desire as theoretical quantities that provide a means of explaining behavior, in particular, betting behavior and other behavior in the face of uncertainty. Psychologists such as Krantz et al. (1971: chaps. 5, 8) and Kahneman and Tversky (1979) present theories of preference and behavior using degrees of belief and desire. These theories are versions of expected utility maximization – in Kahneman and Tversky, a version adjusted to take account of common, systematic

---

[13] Broome (1999: chap. 2) identifies two concepts of utility in contemporary economics. According to one, utility represents preference. According to the other, utility measures the good. Although I adopt neither of these concepts of utility, my concept is more closely related to the first.

[14] I hold that the interval scale for degrees of desire is convertible into a ratio scale by adopting indifference as the zero point, and that ratio comparisons with respect to indifference are justifiably taken as primitive. Nonetheless, I put degrees of desire on an interval scale because interval comparisons are sufficient for the principles governing rational degrees of desire. The additional information given by ratio comparisons is superfluous for expected utility comparisons of options.

departures from straightforward expected utility maximization. Operationist methods of using preferences revealed in behavior to measure degrees of belief and desire, although not definitional for me, assist in their introduction also. The literature on behavior given uncertainty is vast and forms an ample rough introduction of primitive concepts of degree of belief and desire.[15]

I advance a principle of expectation for degrees of belief and desire as well as for probabilities and utilities. Degrees of belief and desire ought to satisfy an expected desirability principle. As I take degrees of belief and desire, they are not necessarily rational. Degrees of belief may violate the basic laws of probability, such as additivity. And degrees of desire may violate the basic laws of utility, such as the expectation principle. I reserve the terms probability and utility for *rational* degrees of belief and desire. More precisely, by probability and utility I mean degrees of belief and desire that are rational and obtain under ideal conditions where excuses for violating the basic laws are absent. Hence probability and utility satisfy the basic laws by necessity, granting that satisfaction of those laws is a requirement of rationality for ideal agents.[16]

When degrees of belief and desire satisfy the basic laws, I call these quantities probabilities and utilities, and I call an option's expected desirability an expected utility. However, I do not assume that the degree of desire for an option is a rational degree of desire, or a utility properly so called. I argue that it is if it equals the option's expected

---

[15] See Weirich (1977) for a detailed presentation of degree of belief and desire, and a detailed introduction of probability and utility. Consider also Jeffrey's (1992: 29) view that degrees of belief and desire are quantitative representations of belief and desire states that are not necessarily quantitative themselves.

[16] I accept the basic laws of probability as supported by intuition and not in need of independent support. Nonetheless, here is a sketch of a method of justifying the laws of probability for degrees of belief, a type of calibration argument, an alternative to the Dutch Book Argument. Consider the epistemic goal for beliefs: truth and avoidance of error. Belief in a contradiction is irrational because it guarantees frustration of the goal. A similar epistemic goal for degrees of belief is to match objective probabilities, as principles of direct inference require when objective probabilities are known. Objective probabilities follow the laws of probability, so degrees of belief must too, or the goal is sure to be frustrated. Moreover, a subsidiary epistemic goal for degrees of belief, given uncertainty about objective probabilities, is to match strength of evidence, and, arguably, strength of evidence follows the laws of probability. So degrees of belief must follow the laws of probability or this subsidiary goal is also sure to be frustrated. Ramsey (1931: 187–8), van Fraassen (1983), and Joyce (1998) provide precedents for this line of argument.

desirability, or expected utility. For convenience, I sometimes call mistaken degrees of belief and desire probabilities and utilities despite their failure to satisfy the basic laws of probability and utility. For instance, when I consider whether the expected utility law is correct, I sometimes ask whether rationality requires that *utilities* satisfy the law. This is intended only as a brief way of asking whether rationality requires that *degrees of desire* satisfy the expectation principle given that conditions are ideal. If the expected utility law is correct, utilities satisfy the law by necessity.

Some authors, such as Davidson (1980: 236–8), claim that a certain amount of rationality in belief and desire is needed to make sense of degrees of belief and desire. One version of this view holds that if a person has intransitive preferences, it is impossible to assign degrees of desire to the objects of those preferences; no assignment represents those preferences.[17] This point assumes, however, that degrees of desire are defined so that they represent preferences. This assumption is common in the operationist school, but the contextualist understanding of degrees of belief and desire dispenses with it. Degrees of desire are assumed to represent only strengths of desire. The preferences of a fully rational person in ideal conditions go by strengths of desire, but the preferences of an irrational person may fail to go by strengths of desire and so be intransitive. Strengths of desires, although transitive, may coexist with preferences that are intransitive, if, irrationally, preferences do not agree with strengths of desire. Also, the strength of a desire depends on quantitative comparison to the desire for the truth of a particular proposition used to establish the unit for strengths of desire relative to a conventional zero point. It does not depend on a function from objects of desire to numbers that represent preferences. The comparisons yielding strengths of desire imply preferences with respect to the proposition establishing the unit, but not with respect to other propositions. So preferences involving the other propositions may be intransitive.[18]

---

[17] Some recent articles question whether rationality requires transitivity of preferences. See, for instance, Sobel (1997). I assume that some version of the principle of transitivity holds. Details are not important here because I use the principle only for the sake of illustration.

[18] The incoherence of intransitive preferences manifests itself in strengths of desire in some ways. Strengths of desire are relative to a unit proposition. Changing the unit generates new strengths of desire. They may not be positive linear transformations of the original strengths. For example, if an agent prefers a to b, b to c, and

It is ironic that taking degrees of belief and desire as primitives yields concepts that are more realistic in some respects than their operationist counterparts. Degrees of belief and desire in my sense do not necessarily satisfy the stringent laws imposed on their operationist counterparts. They are psychological quantities, not constructs for the mathematical representation of preferences. Hence they may exist in cases where the agent's mental attitudes are not structured in the way required for the existence of their operationist counterparts.

3.2.5. Expected Utility Analysis Given Independence

Expected utility analysis claims that the utility of an option $U(o)$ is the expected utility of its possible outcomes, a probability-weighted average of the utilities of its possible outcomes. Given independence of states from options,

$$U(o) = \Sigma_i P(s_i) U(o \text{ given } s_i),$$

where Σ_i stands for summation with respect to the index i, s_i ranges over the members of a partition of states, and $U(o \text{ given } s_i)$ represents the utility of the outcome of o given s_i. Section 4.2 removes the restriction that states are independent of options.[19]

The principle of pros and cons justifies my version of expected utility analysis. The pros and cons are chances for sets of possible worlds

---

c to a, then with a as unit the strength of desire for c is greater than the strength of desire for b. But with b as unit, the strength of desire for c is less than the strength of desire for b. Although the two scales for strength of desire do not cohere, strengths of desire still exist relative to each. They rest on comparison of desires with the desire for the unit proposition.

One scale may better represent strengths of desire than others. It may better reflect the agent's stable goals. Then typically preferences not agreeing with strengths of desire relative to it are irrational and should change to achieve transitivity. For example, given two laundry detergents a shopper may generally prefer the one with the lower price. But in a moment of distraction he may prefer the one in the more colorful box and thereby make his preferences intransitive. Then the scale for strengths of desire that best follows price has a privileged status. Preferences not in alignment with it, such as the one triggered by box color, should change to restore transitivity. Strengths of desire relative to a privileged scale may be called strengths of desire *tout court*.

[19] Applying expected utility analysis using infinite partitions, and to cases where utilities have infinite values, calls for calculus and nonstandard numerical analysis in the style of Robinson (1966). See Sobel (1996) and Vallentyne and Kagan (1997: 7–9) for some discussion of applications of nonstandard numerical analysis to utility theory.

under my interpretation of outcomes. Because outcomes are generated by a partition of states, they do not omit or double-count any relevant consideration. Every relevant consideration has some influence on the utility of some chance, and no relevant consideration exerts its influence twice.

Expected utility analysis may be applied with respect to conditions. Suppose, for example, that a decision tree is constructed for some decision problem, as in Figure 3.1. The decision node generates option branches. Some option branches lead to chance nodes generating branches that represent a partition of states. The tree's paths have terminal nodes that represent the outcomes of options, or the outcomes of options given states. Then suppose that an option-state path is expanded by replacing its terminal node with a chance node and a set of branches representing a partition relative to the state, as in Figure 3.2. The sum of probability-utility products for the possible outcomes issuing from that new chance node is the expected utility of the option given the state leading to the old terminal node. The utility of the option's outcome at the old terminal node equals the expected utility of the option given that state. That is, the utility at the old terminal node equals the expected conditional utility at the new chance node.

As illustrated, expected utility analysis applies to terminal nodes of decision trees, or to options given states. The expected utility of an option o given a state s uses a set of states $\{t_j\}$ that is a partition given s. The probabilities used by the expected conditional utility analysis are the probabilities of these states given s. The utilities used are the utilities of the option's outcomes in these states given s. In general, the utility of an option given a condition is its expected utility given the condition. Under this section's assumption that states are independent of options,

$$U(o \text{ given } s) = \Sigma_j P(t_j \text{ given } s) U(o \text{ given } (t_j \text{ and } s)).$$

My version of expected utility analysis is general in the sense of applying conditionally as well as nonconditionally.

Expected conditional utility analysis creates additional ways of computing an option's expected utility. Given the principle of pros and cons, an option's utility is invariant with respect to trees. So an option's expected utility is invariant with respect to all methods of computation. Its value computed using a partition is the same no matter whether a utility for an option-state pair is assessed directly or is com-

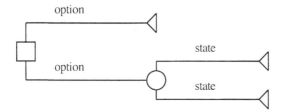

Figure 3.1. A Decision Tree

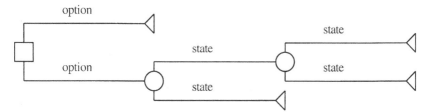

Figure 3.2. An Expanded Decision Tree

puted by expected conditional utility analysis using a partition given the state. Section A.2 shows how to verify this. It provides a consistency check of my applications of the principle of pros and cons, although some details are left as an exercise.

3.A. APPENDIX: REPRESENTATION THEOREMS

This appendix discusses a technical topic bearing on expected utility analysis, namely, representation theorems. It explains why I do not place such theorems at the center of my decision theory.

3.A.1. Definition

Representation theorems are prominent in contemporary decision theory. They show how to extract an agent's probability and utility assignments from her preference ranking of gambles, given that her preference ranking obeys certain axioms. The axioms are motivated chiefly by considerations of rationality. A typical axiom asserts the transitivity of preferences: if gamble a is preferred to gamble b, and gamble b is preferred to gamble c, then gamble a is preferred to gamble c. But some axioms are motivated by technical considerations. Their satisfaction guarantees that preferences have a rich structure. A typical axiom of this type asserts the completeness of the agent's preference ranking of an infinite set of gambles, those generated by events to which probabilities are to be assigned: for every pair of gambles in the set, the agent either prefers one to the other, or is indifferent between them. A representation theorem shows that given satisfaction of the axioms, (1) there are probability and utility functions that agree with the preference ranking in the sense that one gamble's expected utility is at least as great as another's if and only if the first gamble is at least as high as the second in the preference ranking, and (2) these probability and utility functions are unique, if we put aside scale transformations.[20]

The operationist argues for defining probability and utility as the functions whose existence and uniqueness are proved by the representation theorems. He argues that by defining probability and utility in terms of preferences in this manner, probability and utility are made clear – the quantitative is reduced to the comparative.

Some immediate difficulties with the operationist account of quantities are noteworthy. Strictly speaking, operationism must hold that quantities are meaningless until defined in terms of observables. It requires definition, not definability, for meaningfulness. So, strictly speaking, operationism holds, for instance, that quantitative weight was meaningless before Hölder's (1901) representation theorem provided a definition in terms of comparative relations. Furthermore, because definitions of a quantity based on different representation theorems yield different concepts even if the definitions are logically equivalent, operationism holds that no single concept of weight arises from the different representation theorems for weight.

It is also worth noting that defining probability and utility via the representation theorems compromises the operationist's standards to some extent. The operationist requires definition in terms of things directly and intersubjectively observable and not merely inferable by induction from things directly and intersubjectively observable. But

---

[20] Von Neumann and Morgenstern (1944: sec. 3) state a well-known representation theorem for utility, Savage (1954: chaps. 3, 5) proves representation theorems for both probability and utility, Jeffrey (1983: chaps. 8, 9) proves a representation theorem for probability and utility taken as functions of propositions, Maher (1993: chap. 8) proves a representation theorem for cognitive utility, and Joyce (1999: chap. 4) proves a representation theorem generalizing evidential and causal decision theory.

definitions of probability and utility via the representation theorems do not limit themselves to observables of this type. First, preferences, the starting points of the theorems, are not directly and intersubjectively observable. Nor are the choices they prompt. Only the overt behavior that the preferences and choices cause is directly and intersubjectively observable. Preferences and choices, because they have a mental aspect, are inferred from overt behavior and not definable in terms of it. Second, the representation theorems generally require an infinite number of preferences, more than any person has. Often they are applied to hypothetical preferences, preferences that a person would have in certain hypothetical situations. These hypothetical preferences, expressed in terms of counterfactual conditionals, are not definable in terms of observables. The counterfactual conditional connective does not stand for an observable, nor is it definable in terms of observables. Unless operationism licenses it as an admissible auxiliary expression, along with logical vocabulary, it does not belong in operational definitions. For these reasons, the definitions of probability and utility via the representation theorems do not fully achieve the goal of operationism. Operationists in decision theory have been inclined to accept the ways in which the definitions fall short of their goal because they want to make advances in theory construction. In order to construct a theory of decision, they have been willing to settle for the limited progress toward the observable that comes from defining probability and utility in terms of hypothetical preferences.

Some operationists intensify the need for compromise by using choice to define preference and then preference to define utility. This is the method of revealed preference theory. The definitions require an infinite set of hypothetical choices, and so go a step further from the operational than an infinite set of hypothetical preferences. Actual choices at a moment are too few. And using actual choices across a period of time also does not provide enough choices and in addition creates an unmet demand for an operationist means of distinguishing inconsistent choices from choices prompted by changing preferences.

Binmore (1998: 362) addresses the problem. He says that revealed preference theory assumes that the choices used to define preferences are made during a period in which the underlying determinants of an agent's choice behavior are fixed. But there is no operational test for the constancy of the underlying determinants of behavior. Furthermore, because preference in the ordinary sense is one of the determinants of choice behavior, revealed preference theory is circular if it attempts a definition of preference in the ordinary sense. To prevent a circle, the theory must settle for an explication of preference rather than a definition. That is, the theory must propose an operationist substitute for preference in the ordinary sense. The technical substitute may disagree with preference in the ordinary sense and, in any case, lacks ordinary preference's explanatory power, as Binmore (1998: 519) admits. This is its drawback from a theoretical perspective.

In contrast with operationism, contextualism in decision theory accepts probability and utility as primitives introduced via the theory of probability and utility, not defined in terms of comparative relations via representation theorems. Contextualism does not discard the representation theorems, however. It reinterprets their significance for decision theory. Even if they do not define probabilities and utilities, they provide a means of deductively inferring probabilities and utilities from preferences, that is, a means of measuring probabilities and utilities. They show that when the axioms of preference are satisfied, probabilities and utilities may be inferred from preferences, under the assumption that preferences maximize expected utility. The values of probabilities and utilities are given by the functions whose existence and uniqueness are asserted in the representation theorems.[21]

Casting the theorems as tools of measurement sufficiently accommodates the methodological constraints of the empirical sciences. The sciences' testability requirements are met if probabilities and utilities may be inferred from observables, even if they are not defined in terms of observables. The testability requirements do not force the sciences to define probability and utility via the representation theorems. Restricting decision theory to terms operationally defined is insufficiently motivated, even for the sciences.

To support a contextualist interpretation of the representation theorems, I point out the heavy cost of defining probability and utility via the representation theorems. Given their definition via the representation theorems, it is impossible to provide a general justification of preferences by showing that preferences maximize expected utility. For if one defines probabilities and utilities so that preferences maximize expected utility, one cannot then justify preferences by showing that they maximize expected utility. The most one can do is to define some

---

[21] See Sobel (1998) for a discussion of the role of representation theorems in decision theory.

probabilities and utilities in terms of a selective base set of preferences, and then use those probabilities and utilities, and the expected utility rule, to justify preferences outside the base set. It is impossible to justify all preferences by means of probabilities and utilities, and the expected utility rule, because some preferences must be taken for granted to define probabilities and utilities. Operationist decision theory is thus unable to provide a general justification of preferences in terms of expected utilities.[22]

An analogy may clarify the point. Consider the definition: a bachelor is an unmarried man. To explain why a person is a bachelor, it does no good to say he is an unmarried man. Because being a bachelor entails being an unmarried man by definition, one needs to say more. Perhaps the explanation of the man's bachelorhood is disinclination to live with others. Similarly, to explain the rationality of a preference, it does no good to say that the preference maximizes expected utility if expected utilities are defined so that they represent preferences. Expected utilities justify preferences only if they represent facts definitionally independent of those preferences.

Another drawback arises even in cases where certain preferences are justified by means of probabilities and utilities defined in terms of a base set of preferences. The definition of probabilities and utilities requires that the preferences in the base set meet the axioms of preference, and these axioms are demanding. They are rarely met in any actual case. Their satisfaction is usually taken as a convenient simplifying assumption. Now justifications that employ simplifying assumptions may be valuable. But the stronger the simplifying assumptions, the less valuable the justifications. Dispensing with the simplifying assumption that a base set of preferences meets the axioms would make the justification of preferences outside the base set more valuable. If probabilities and utilities are taken as primitives, justifications of preferences may dispense with the simplifying assumption that the preferences in a base set satisfy the axioms. The justifications may go directly from probabilities and utilities to preferences without the mediation of a rich and highly structured base set of preferences. As a result, expected utility justifications of preferences are more realistic.

---

[22] Hampton (1994: 207–10) observes that a representation theorem such as von Neumann and Morgenstern's merely establishes that preferences satisfying certain axioms can be represented as maximizing expected utility. She says that this result has no normative significance.

For example, suppose that a person's preferences are riddled with intransitivities so that there is no base set of preferences in terms of which probability and utility can be defined. The operationist then cannot justify any of the person's preferences in terms of an actual base set of preferences. But the contextualistic can justify the person's preference for a bet on the Cubs over the Cardinals, for instance, if the bet maximizes expected utility, given the person's subjective probabilities that the teams will win. Intransitivities in his preference structure do not infect the justification of his preference taken in isolation.

In general, instead of justifying some preferences in terms of others, a contextualist decision theory justifies preferences in terms of primitive probabilities and utilities. It uses a principle of expected utility that is completely general and not restricted to cases where structural axioms for preferences are satisfied. It does not take the principle of expected utility to be satisfied by definition, but supported by argument and intuition, and derived from more fundamental principles of utility, as in Section 3.1.

The dispute between the operationist and the contextualist about the interpretation of the representation theorems is basically a dispute about the trade-off between clarity of concepts and quality of explanations. The operationist opts for clarity of concepts at the cost of quality of explanations. The contextualist opts for quality of explanations at the cost of clarity of concepts. Since the standards of clarity the operationist propounds are unjustifiably strict, the contextualist's compromise is wiser.[23]

3.A.2. Cognitive Power

Some decision theorists hold that in an ideal case where the decision maker has unlimited cognitive power and can think without cost, probabilities and utilities will be formed, so that an idealization about their availability is redundant. The unavailability of probabilities and utilities in real-life cases, they claim, is a result of not thinking hard enough.

[23] Christensen (1996: 453–4) makes kindred points about operationism during an examination of the Dutch Book Argument for the laws of probability. He thinks that many versions of the argument are marred by an operationistic account of degree of belief. He seeks a nonoperationistic version of the argument. Joyce (1999: 19–22) abandons behaviorism, a method allied with operationism, for reasons similar to my objections to operationism. He advocates basing decision theory on desires, not overt actions, to escape the narrow confines of behavioristic definitions.

If one fully formed preferences concerning gambles, one could use representation theorems to infer probabilities and utilities. This view about the consequences of cognitive power is mistaken, however. When information is short, the formation of the preferences necessary for applying representation theorems is unwarranted. Just as one should suspend belief when information is insufficient to justify a belief, one should suspend the formation of preferences when information is insufficient to justify them, and then probabilities and utilities may be indeterminate. So I take the formation of probabilities and utilities as an independent component of my large idealization about agents.[24]

3.A.3. Justification

Resnik (1987: 99), Joyce (1999: 79–82), and others claim that decision theory's representation theorems furnish a justification for expected utility analysis. There are two ideas behind the claim. The first idea is that if the expected utility principle for preferences is correct, and rationality requires preferences maximizing expected utility, then expected utilities are utilities as expected utility analysis asserts. The second idea is that a representation theorem shows that, given satisfaction of its structural axioms of preference, satisfaction of the remaining axioms of preference – principles of rationality – ensures that preferences agree with expected utilities, that is, comply with the expected utility principle. I have no quarrel with the first idea but the second, even if true, does not justify the expected utility principle. A representation theorem shows that if a person's preferences satisfy certain structural conditions and certain plausible rationality

---

[24] Resnik (1987: 24) holds that a complete preference ranking of gambles is a goal of rationality, one met by rational ideal agents in ideal circumstances. Such a complete preference ranking induces quantitative beliefs and desires. Hence given Resnik's position, and granting that a rational agent's cognitive goals include having complete probability and utility assignments, and that no standard of rationality prohibits the formation of quantitative probability and utility assignments, an ideal agent in ideal circumstances has complete probability and utility assignments if fully rational. This case for complete probability and utility assignments is not compelling. Resnik's assumption is questionable. A complete preference ranking of gambles does not seem to be a goal of rationality. Rationality is content with incomplete preferences when information is short. In fact, rationality may prohibit preference formation in the face of an information shortage. See Kaplan (1996: chap. 1, sec. V) and Schick (1997: 45–7) for additional argumentation that rationality does not demand a complete preference ranking.

conditions, then there are probability and utility functions over the outcomes of options such that the expected utilities of options according to those functions agree with preferences between options. The theorem merely points out that an expected utility representation of preferences is possible, given satisfaction of the structural and rationality conditions. If, according to operationism, probability and utility are defined as the functions that yield that expected utility representation, then compliance with the expected utility principle follows by definition. Compliance is not justified by the rationality conditions. On the other hand, if probability and utility are taken as primitives, in accordance with contextualism, then the theorem does not guarantee that the expected utility principle is satisfied when preferences meet the axioms, only that its satisfaction is compatible with the preferences. Either way the theorem does not justify the expected utility principle and so does not justify expected utility analysis.

Take, for example, a simple case in which the agent forms preferences between bets that pay $1 or $0 depending on whether the proposition bet upon is true or false. Let a bet that p be represented by $B(p)$, let ">" stand for preference, and let "$\approx$" stand for indifference. Suppose that preferences are as follows: $B(p \vee \sim p) > B(\sim p) \approx B(p) > B(p \& \sim p)$. These preferences satisfy all the rationality axioms of the representation theorems, and they satisfy the structural axioms of the representation theorems for finite preference orderings. Suppose that one defines probability and utility as the functions, unique up to scale transformations, that yield expected utilities agreeing with these preferences. Then adopting certainty that a proposition is true as the unit for probability, and adopting certainty that a proposition is false as the zero point for probability, $P(p) = P(\sim p) = .5$. Also, setting the unit and zero point for utility so that $U(\$1) = 1$ and $U(\$0) = 0$, $U(B(p)) = U(B(\sim p)) = .5$. Under this probability and utility assignment, preferences maximize expected utility, but this is no justification of the expected utility principle. If one does not define probability and utility so that the principle is satisfied, the preferences may turn out to violate the expected utility principle. For instance, if the utilities of $1 and $0 are as given, and the probability of p is .4 and the probability of $\sim p$ is .6, then the preferences violate the expected utility principle. The principle says that $B(\sim p)$ should be preferred to $B(p)$, whereas the agent is indifferent between the two bets.

The representation theorems permit successful inference of probabilities and utilities only when the agent is ideal and has completely rational preferences, preferences conforming to expected utilities. The idealization required makes it clear that the representation theorems do not justify the expected utility principle. Their application assumes the expected utility principle rather than justifies it.

The rationality conditions, such as transitivity of preference, invoked by the representation theorems are too weak to justify the expected utility principle. They are just necessary conditions for satisfaction of the principle. Collectively they are not much more plausible than the expected utility principle itself. In fact, the main reason for accepting some is that they are necessary conditions for satisfaction of the expected utility principle. Their appeal is mainly that they are weaker than the expected utility principle. They are the comparative counterparts of the quantitative expected utility principle.

The most the representation theorems can do is partially justify the expected utility principle. Suppose that preferences satisfy the structural and rationality axioms so that the probability and utility functions that yield expected utilities agreeing with the preferences are unique aside from scale transformations. And suppose that preferences do maximize expected utility so that the functions specify genuine probabilities and utilities. Then suppose that a new preference is formed without introducing a violation of the structural axioms. If the enlarged set of preferences also satisfies the rationality axioms, then the new preference does not violate the expected utility principle. That is, once expected utility maximizing preferences are rich enough to provide a means of inferring probabilities and utilities, new preferences that do not violate the structural axioms must comply with the expected utility principle or else violate the rationality axioms.

To see this, suppose that the enlarged set of preferences meets the structural and rationality axioms. Then there is just one pair of probability and utility functions unique up to scale transformations that yields expected utilities agreeing with preferences in that set. These functions must agree with the original probability and utility functions. That is, if a preference maximizes expected utility according to the new probability and utility functions, it cannot fail to maximize expected utility according to the original functions, which by assumption give genuine expected utilities. Because the new preference maximizes expected utility according to the new probability and utility functions, it does not fail to maximize genuine expected utility.

For example, suppose an agent prefers p to q and is indifferent between a gamble that yields p if s and q otherwise, and a gamble that yields p if $\sim s$ and q otherwise. Then, given the assumptions, he assigns s probability .5. So if he prefers p' to q' and is offered analogous gambles on s concerning them, he must go by expected utilities and be indifferent between the gambles, or else violate an axiom for rational preference rankings. This partial justification of the expected utility principle falls short of a full justification because it assumes that the base set of rational preferences satisfies the expected utility principle, and so argues for the principle only in extensions of preferences beyond the base set. It does not argue for satisfying the expected utility principle in the base set of preferences.

Because representation theorems are not a central feature of my contextualistic theory of decision, I do not formulate a representation theorem for my version of expected utility analysis. Let me note, however, that my version falls into the camp of causal decision theory (see Section 4.2). Representation theorems have been established for versions of expected utility analysis in this camp. See, for example, Armendt (1986, 1988) and Joyce (1999: chap. 7). The methods of proof should carry over to my version.

4

Expected Utility's Promotion

The previous chapter presented the main ideas of expected utility analysis. This chapter defends those ideas against objections and generalizes them for cases where options and states are not independent.

One of the main objections to expected utility analysis challenges the assumption that a partition of states yields a suitable separation of pros and cons. The objection claims that a partition of states separates pros and cons in a way that omits some relevant considerations, in particular, risk. Here by risk I mean the epistemic possibility of gain or loss. This is the ordinary meaning enlarged to include the possibility of gain as well as loss, and refined to take risk specifically as an epistemic rather than a physical phenomenon. The objection claims that the risk involved in an option depends on the set of possible outcomes and does not appear in the possible outcomes themselves. As a result, expected utility calculations ignore some considerations bearing on an option's utility.

Allais's paradox (1953) clearly displays this line of argument. Simplified, the paradox goes as follows. It is not irrational (a) to prefer $3,000 for sure to a 4/5 chance for $4,000, and simultaneously (b) to prefer a 1/5 chance for $4,000 to a 1/4 chance for $3,000. But no utility function U exists such that (a) $U(\$3,000) > 4/5\ U(\$4,000)$ and (b) $1/5\ U(\$4,000) > 1/4\ U(\$3,000)$, as we see plainly after multiplying both sides of the second inequality by 4. The reason the preferences are not irrational, despite the apparent disregard for expected utilities, is related to risk. Risk is at least partly a matter of the dispersion of the utilities of possible consequences. In the second preference the dispersion is about the same for both options. Aversion to risk does not strongly favor either option. The preference depends primarily on the utilities

117

of the chances for the monetary gains. But in the first preference one option, the sure thing, has no dispersion and so no risk. That option enjoys a special advantage; it does not incite aversion to risk. This special advantage accounts for the first preference. The two preferences are not irrational if one considers risk. According to the objection, the conflict between preferences and an expected utility ranking of options arises because the utilities of the possible monetary gains used in expected utility analysis do not take account of the risks involved in the options. Expected utility analysis in terms of monetary gains fails because it omits an important consideration.[1]

Ellsberg's paradox (1961) also illustrates the point. Simplified, the paradox goes as follows. Each of two urns has a mixture of red and black balls. The mixture in the first urn is known to be 50–50. The mixture in the second urn is completely unknown. It is not irrational (a) to prefer a gamble that pays $100 if a red ball is drawn from the 50–50 urn to a gamble that pays $100 if a red ball is drawn from the mystery urn, and simultaneously (b) to prefer a gamble that pays $100 if a black ball is drawn from the 50–50 urn to a gamble that pays $100 if a black ball is drawn from the mystery urn. However, no utility function U exists such that (a) $P(R1)U(\$100) > P(R2)U(\$100)$ and (b) $P(B1)U(\$100) > P(B2)U(\$100)$, as we see plainly after adding the two inequalities and recalling that $P(R1) + P(B1) = P(R2) + P(B2) = 1$. The reason the preferences are not irrational, despite the apparent disregard for expected utilities, is related to risk again. This time, however, the relevant risks are generated by lack of information about the mixture in the mystery urn. Because of that lack of information, the gambles concerning the mystery urn are riskier than their counterparts concerning the 50–50 urn. Aversion to risk accounts for the preferences for the gambles concerning the 50–50 urn. Again, according to the objection, expected utility analysis in terms of monetary gains omits a crucial consideration.[2]

---

[1] See Machina (1982) for a revision of expected utility analysis prompted by Allais's paradox. The paradox raises primarily a problem concerning risk, I think. But some theorists hold that regret is the central issue. They claim that expected utility analysis survives confrontation with the paradox if an option's outcome covers regret for missed opportunities. See Weber (1998) for a critical discussion of that defense of expected utility analysis.

[2] Cases such as Ellsberg's lead Gärdenfors and Sahlin (1982) to modify the decision rule to maximize expected utility. They change it to accommodate risks generated by indeterminate probabilities. For a discussion of their proposal, see Weirich (2001).

The standard operationistic definition of aversion to risk takes it as the concavity of a utility function derived from preferences between gambles via a representation theorem.[3] According to such a definition, aversion to risk does not explain the paradoxical preferences in the foregoing examples, because no utility function of whatever shape accommodates those preferences. Operationists are inclined to declare the paradoxical preferences irrational, not recognizing the kind of aversion to risk that explains them. My contextualist account of aversion to risk, however, provides for the legitimacy of those preferences. I have to defend expected utility analysis against the objection they raise.

The objection can be circumvented by circumspection. It assumes a narrow view of outcomes. It does not count the risk involved in an option as part of the outcome of the option. In fact, in the foregoing examples it takes monetary gains as the only relevant outcomes of options. The apparent conflict between preferences and expected utility principles in the examples disappears if risk is included in the outcomes of the options. If one assumes an aversion to risk, in Allais's paradox the outcome of $3,000 for sure enjoys an advantage since that option involves no risk. Similarly, in Ellsberg's paradox the outcomes of gambles on the 50–50 urn enjoy an advantage because those gambles involve less risk than their counterparts with the mystery urn. These advantages concerning outcomes are sufficient to bring expected utilities into agreement with preferences. In general, expected utility analysis does not fail to take account of risk. Risk is part of the outcome of a risky option; it is part of each of the risky option's possible outcomes, a sure consequence of the option. My version of expected utility analysis, which takes an option's outcome given a state comprehensively, puts risk in each possible outcome of a risky option and so does not omit that relevant consideration.[4]

---

[3] Pratt (1964) and Arrow (1965: 33; 1970) propose defining aversion to risk as the concavity of the utility function for a commodity. Their operationistic definition of aversion to risk has some drawbacks as a definition of the ordinary concept of aversion to risk. First, it does not rest on a definition or introduction of risk itself. Second, it conflates the diminishing marginal utility of a commodity with aversion to risks concerning the commodity. Third, it requires that aversion to risk be commodity relative. And fourth, it does not provide for a particular amount of a commodity figuring as a possible consequence of several options of varying degrees of risk. See Weirich (1986) and Bell and Raiffa (1988).

[4] Taking risk as a consequence of gambles also provides a way of resolving the St. Petersburg paradox. See Weirich (1984b). The main idea is that aversion to the risk

Does putting risk into outcomes defeat the purpose of expected utility analysis? Isn't its purpose the separation of probabilistic and utilitarian considerations? Shouldn't outcomes involve only utilitarian considerations? As I conceive of it, expected utility analysis's purpose is the computation of an option's utility. It does this by separating pros and cons according to possible outcomes. Its success does not depend on a separation of probabilistic and utilitarian considerations. As long as all relevant considerations are counted exactly once, and given their proper weight, it works. Packing some probabilistic considerations into outcomes is not contrary to its design. In fact, an expected utility analysis separates probabilistic and utilitarian considerations concerning gambles only if (1) it uses possible worlds as outcomes and (2) none of the agent's basic intrinsic attitudes concerns risk or other factors with a probabilistic component. If an agent has a basic intrinsic aversion to risk, probability and utility are inseparably intertwined. If an analysis uses coarse-grained outcomes, their utilities depend on the probabilities of their finer-grained realizations. As Section A.2 shows, the dependency ensures that the coarse-grained analysis yields the same result as any finer-grained analysis. Textbook examples may create the impression that the purpose of expected utility analysis is to separate probabilistic and utilitarian considerations. They typically focus on the probabilities of the results of games of chance and the monetary outcomes of bets on those games. Outside the textbook, however, an expected utility analysis often uses coarse-grained outcomes whose utilities involve probabilistic considerations.

Some may object that taking outcomes broadly, as I do, makes expected utility analysis nonfalsifiable, and so empty. They may say that one can always introduce a special outcome like risk to save expected utility analysis from alleged counterexamples. But the possibility of introducing a special outcome to save the method of analysis does not make it nonfalsifiable. To stave off falsification, the special outcome has to be relevant, one that the agent cares about. It is not true that taking outcomes broadly one can always introduce a special relevant outcome to save expected utility analysis. A putative counterexample can begin by specifying the relevant outcomes. If rational preferences can be brought into conformity with expected utility analysis only by

involved in the St. Petersburg gamble reduces the utility of its possible outcomes to the point that the utility of the gamble is finite, although its expected monetary value is infinite.

introducing outcomes that are irrelevant, then the counterexample is genuine.[5] In Allais's and Ellsberg's paradoxes, I could have specified that the only relevant outcomes are monetary gains. Then introducing risk to make expected utility analysis accommodate the preferences would have been illegitimate. If the preferences are not irrational, given that the only relevant outcomes are monetary gains, then expected utility analysis is refuted (if preferences agree with utilities). Expected utility analysis survives, however, because under the revision about what matters the preferences are irrational.[6]

Moreover, the objection about nonfalsifiability assumes that the only falsifying data are rational preferences. It assumes that degrees of belief and desire are mere representations of preferences. Taking degrees of belief and desire as primitive enlarges the scope of falsifying data. Rational degrees of belief and desire that violate the expected utility principle are falsifying data. The broad view of outcomes in no way rules out such data. Given degrees of belief and desire attaching to broad outcomes and violating the expected utility principle, one cannot make the violation disappear by changing the scope of outcomes, in particular, by making outcomes broader. The expected utility principle as I have formulated it uses comprehensive outcomes that are maximally broad and so cannot be made broader.

Some may fear that risk is double-counted if it is counted as an outcome. They may think that the calculation of expected utility is itself a means of giving risk its due. A review of expected utility analysis from the perspective of the principle of pros and cons dispels this worry. In an expected utility analysis of an option's utility, the only

---

[5] Broome (1991: 103) defines a "justifier" as a difference between outcomes that makes it rational to have a preference between them. A list of justifiers may replace a list of relevant outcomes as a means of providing for counterexamples to expected utility analysis. If the only way to bring rational preferences into alignment with expected utility analysis is to take the utility of outcomes as sensitive to nonjustifiers, then the counterexample is genuine. A list of justifiers is more controversial than a list of relevant outcomes, however, because the former is normative whereas the latter is descriptive.
    Broome (1999: chap. 5) claims that because utility theory needs justifiers to make it nonempty, instrumental theories of rationality may not adopt the consistency or coherence requirements utility theory advances. This argument presumes that basic intrinsic attitudes may not function as justifiers. It also saddles instrumentalism with an operationist account of preference.

[6] Raiffa's (1968: 80–6) defense of expected utility analysis against Allais's paradox claims that the paradoxical preferences are irrational. His defense assumes that monetary gains are the only relevant outcomes.

considerations acknowledged are chances for possible outcomes. These chances are assigned utilities, and the utilities are added to obtain the option's utility. The analysis does not take account of risk unless risk is included in the possible outcomes. Its inclusion does not result in double-counting.

Risk is involved in dependency relationships that may create the appearance that its inclusion in outcomes leads to double-counting. The risk involved in an option depends on the option's generation of chances for possible outcomes. If that risk is included among the possible outcomes themselves, then it depends in part on itself. However, this feedback relationship does not cause a double-counting of risk. Although the risk involved in a possible outcome depends on the possible outcomes, it is not double-counted when the utilities of chances for possible outcomes are summed. It is counted only as a component of the possible outcome and not a second time as a cause of the possible outcome. By way of analogy, consider a distribution of interpersonal utility among people who care about the equality of the distribution. Although the utility of the distribution for a person depends on its equality, the utility generated by its equality is not double-counted when the utilities it generates for individuals are added. Its equality is counted only as a component of a person's utility and not a second time as a cause of that utility.

The main objection to including risk in outcomes is the operationist objection that including risk makes it harder to derive outcomes' probabilities and utilities from preferences among options. In fact, most representation theorems incorporate some device for excluding risk from outcomes. Savage (1954: 14, 25), for example, excludes risk by requiring that outcomes be general, that is, producible with arbitrary probabilities by many acts, including sure things. Because risk cannot be part of an outcome that is a sure thing, it is excluded.[7]

The operationist objection is sometimes overstated. Although including risk in outcomes complicates applications of expected utility analysis, it does not make expected utility analysis inapplicable by any means. Conclusions about probabilities and utilities still can be derived

---

[7] Axioms such as Savage's limit the application of revealed preference theory. Binmore (1998: 359–62, 392) says that cases where the theory seems to go wrong may be handled by better modeling, that is, for example, by broader descriptions of an act's possible outcomes. But in cases such as Allais's and Ellsberg's paradoxes no modeling improvement works. Making the outcomes as broad as necessary falsifies an axiom of the theory. Choices are rational despite violating the axioms.

from applications of expected utility analysis, as in Weirich (1986: sec. 5). The main reply to the operationist objection, however, concerns its methodological presumptions. The objection may have some force in disciplines where operationalization is necessary for theory testing. But a philosophical decision theory does not face the requirements of observational testability prevalent in the empirical sciences. Expected utility analysis, taken as part of a philosophical decision theory, is immune to the operationist objection. Philosophical decision theory seeks fundamental principles of utility analysis and need not compromise this goal to facilitate empirical applications. If fundamental principles of utility analysis do not lend themselves to operationalization, so be it. We must accept normative decision principles as they are and not as we would like them to be for ease of application.

4.2. GENERALIZATION

Now let me remove the restriction that states are independent of actions, while retaining my idealizations, and formulate expected utility analysis in a more general way. As in Section 3.2, I must specify the chances for possible outcomes whose utilities are added to obtain an option's utility. First, I must specify the possible outcomes in terms of the option and states. Then I must specify each possible outcome's probability and utility; their product yields the utility of the chance for the outcome. As in Section 3.2.3, the outcome of an option-state pair is not taken to be a single possible world but a proposition represented by a set of possible worlds. The proposition involves supposition of the option and the state. To specify an outcome precisely, I must specify the context for the option's and state's supposition, and the ways in which they are supposed. The results must ensure that the set of chances for possible outcomes neither omits nor double-counts any relevant consideration.

4.2.1. Probabilities for the General Case

Suppose, then, that an option has some influence on a state. If the option has desirable consequences given the state and increases the probability of the state, then the option's influence on the state is a positive consideration in favor of the option. On the other hand, if the option has undesirable consequences given the state and increases the probability of the state, then the option's influence on the state is

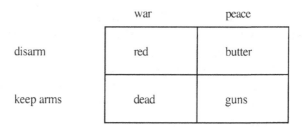

	war	peace
disarm	red	butter
keep arms	dead	guns

Figure 4.1. The Cold War

a negative consideration against the option. If the option decreases the probability of the state, the pros and cons are reversed. Section 3.2's formulation of expected utility analysis overlooks such pros and cons. They do not influence the utilities of chances for possible outcomes, which are added to obtain the option's expected utility. The utilities of chances for possible outcomes are just the utilities of the possible outcomes weighted by their probabilities.

As an example of the oversight, consider the account of the Cold War in Figure 4.1, an account along the lines of an example in Jeffrey (1983: sec. 1.5). The rows represent options and the columns represent states. The boxes represent possible outcomes. Suppose that in case of war being red is preferred to being dead, and if peace prevails having butter is preferred to having guns. Then the outcome of disarming is preferred to the outcome of keeping arms no matter whether there is war or peace. Calculating expected utilities as in Section 3.2, the expected utility of disarming is higher than the expected utility of keeping arms. However, the calculation is wrong because it ignores deterrence, the promotion of peace by keeping arms. I must revise my formulation of expected utility analysis to take account of such considerations.

The obvious remedy is to recalculate the utilities of chances of possible outcomes so that they take account of the influence of options on states. The chance for butter if one disarms has a lower utility than the current probability of peace times the utility of butter. Its utility is really the probability of peace *if one were to disarm* times the utility of butter. In general, when an option influences a state, one should calculate the expected utility of the option using the probability of the state if the option were realized instead of the probability of the state. The formula for expected utility analysis needs $P(s$ given $o)$ in place of $P(s)$. One may use

$P(s)$ to simplify only in cases where s is independent of o so that $P(s) = P(s$ given $o)$.

$P(s$ given $o)$ is my symbol for the probability of s if o were realized. It is a conditional probability assigned during deliberations, not a non-conditional probability assigned at the time of choice upon o's realization, although it may equal that probability. Its expression in terms of suppositions, and its designated role in expected utility analysis, provide an initial introduction. Still, the quantity merits a more complete explanation. What is the probability of a state if an option were realized? Is it equal to other more familiar quantities? The conditional relation involved in $P(s$ given $o)$ is so rich a topic that I cannot do it justice within the confines of this book. I provide only a partial account of $P(s$ given $o)$, focusing on matters crucial for expected utility principles. I assume that the account can be elaborated in a way congenial to the principles.

The handiest interpretation of $P(s$ given $o)$, the probability of s if o were realized, invokes conditional probability as it is understood in probability theory. In probability theory, the probability of s given o, abbreviated $P(s/o)$, is defined as a ratio of nonconditional probabilities, namely, the probability of the conjunction of s and o divided by the probability of o, that is, $P(s \& o)/P(o)$.[8] In the example, the probability of peace if disarmament were chosen is a ratio according to this interpretation. With an obvious scheme of abbreviation, $P(p$ given $d)$ = $P(p/d) = P(p \& d)/P(d)$.

This interpretation faces many problems. One technical problem is that if an agent is certain he will adopt an option o, so that the probability of any other option equals 0, then for any state s and option $o' \neq o$, the ratio $P(s \& o')/P(o')$ involves division by zero so that the conditional probability $P(s/o')$ is undefined, and the expected utility of o' is undefined as well. This is a problem for postdecision evaluations of options more than for applications of expected utility analysis in deliberations prior to a decision, because options of practical interest in deliberations have nonzero probabilities. But even in predecision deliberations the restriction of expected utility analysis to events with nonzero probabilities is unappealing from a theoretical viewpoint. One would like to use the expected utilities of undesirable options to

[8] Do not suppose that $P(s/o)$ is the probability of s if exactly o is learned. The ratio equals the probability of s under the assumption that o. See Eells (1982: 185–6) and Weirich (1983a).

125

explain why they are not contenders. But if an option is not a contender, it lacks an expected utility. The best way of resolving this technical problem is to extend the standard definition of conditional probability to handle conditions with zero probability.[9] I do not work out the details of a resolution, however, because another, insurmountable difficulty prohibits using standard conditional probabilities to define expected utilities.

Newcomb's problem, presented by Nozick (1969), mandates a revised definition of expected utilities. In this decision problem the outcome of each option depends on a prediction about the option selected. The decision maker may take an opaque box that contains $0 or $1 million, or the opaque box together with a transparent box that contains $1,000. The opaque box contains $1 million just in case the predictor anticipated a decision to pick only the opaque box. The predictor is reliable – those who pick only the opaque box usually end up with $1 million, whereas those who pick both boxes usually end up with just $1,000. What is the rational choice: the opaque box or both boxes? Figure 4.2 depicts the options and their possible outcomes.

Picking both boxes is better no matter which prediction has been made. Whatever state obtains, its outcome beats one-boxing's outcome. That is, it strictly dominates one-boxing. When options do not causally influence states, the well-supported principle of dominance says to avoid a strictly dominated option. Since neither one-boxing nor two-boxing causally influences the prediction, which precedes the choice, the principle of dominance recommends two-boxing.

On the other hand, maximizing expected utility according to ordinary conditional probabilities recommends one-boxing. Picking only the opaque box, if that decision is predicted, is better than picking both boxes, if that decision is predicted. Because it is highly probable that one-boxing has been predicted given that it is adopted, maximizing expected utility using ordinary conditional probabilities recommends that option.

Newcomb's problem raises many subtle and intriguing philosophical issues. For an introduction to the range of issues, see Gibbard and Harper (1981), Eells (1982), Campbell and Sowden (1985: pt. 1), Nozick (1993: 41–50), Sobel (1994: chap. 2), and Joyce (1999: chap. 5). I attend to only one issue, however: the divergence of evidential and causal

---

[9] Joyce (1999: 231–2) presents one way of making the extension.

126

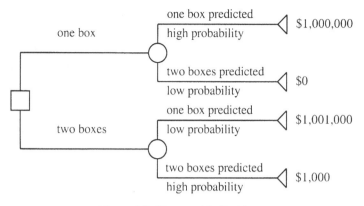

Figure 4.2. Newcomb's Problem

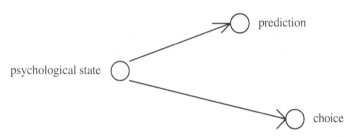

Figure 4.3. The Structure of Newcomb's Problem

relations. To bring out this divergence, I assume that prior to his choice the decision maker is in a psychological state that causes his choice and also its prediction. That is, his choice and its prediction have a common cause. Figure 4.3 depicts the relevant causal and temporal relations. The choice furnishes evidence about the psychological state and thereby evidence about the prediction, but it does not cause the prediction. The correlation between choice and prediction is spurious – that is, not causal but merely evidential. Choosing only the opaque box may produce evidence of a prior disposition to one-box and thereby evidence of a prediction of one-boxing. But it does not cause that prediction. Given the problem's causal structure, one-boxing provides evidence that the opaque box contains $1 million, but it does not affect the contents of the box, which are settled by the prediction made earlier.

Expected utilities computed using ordinary conditional probabilities recommend one-boxing because of the evidence it provides for riches rather than because of its efficacy in producing them. Dominance recommends two-boxing because of its efficacy; it generates $1,000 more than one-boxing, whatever the prediction. Mere auspiciousness is not a good reason for adopting an option. A rational choice aims at efficacy. Hence two-boxing is the rational choice. One-boxing, being less efficacious, is an irrational choice.

Many decision problems, besides Newcomb's problem, offer auspicious but inefficacious options. (1) Suppose that a gene causes both a disposition to smoke and a disposition to contract lung cancer, but that smoking does not actually cause lung cancer. Then not smoking provides evidence that one lacks the gene and thereby evidence that one lacks the disposition to contract lung cancer. But not smoking does not remove the gene or the disposition to contract lung cancer. Although it is auspicious, it is not efficacious. (2) Suppose that smiling at the office is evidence that one's spouse is happy but does not cause that happiness. Then in regard to one's spouse's state of mind, smiling at the office is auspicious but not efficacious. (3) Suppose that being prosperous is evidence of being among the elect but does not cause election, which is predetermined by God. Then in regard to salvation, achieving prosperity is auspicious but not efficacious. (4) Suppose that one's voting is evidence that others, being like-minded, will vote as well, although one's voting does not cause them to vote. Then in regard to voter turnout, one's voting is auspicious but not efficacious. In all these cases auspiciousness is a poor reason to adopt an option.

Newcomb's problem assumes that expected utility is computed using ordinary conditional probabilities. It shows that expected utility then measures an option's auspiciousness rather than its efficacy. This happens because, as is well known in probability theory, standard conditional probability measures evidential support, not causal influence. That is, if the conditional probability $P(s/o)$ is greater than the nonconditional probability $P(s)$, then o's realization furnishes evidence that s but does not necessarily influence s's realization.[10]

---

[10] To see that conditional probability does not measure causal influence, observe that $P(s/o) = P(s)$ if and only if $P(o/s) = P(o)$, if we assume all quantities exist. Thus, if conditional probability measures causal influence, s's realization does not influence o's realization if and only if o's realization does not influence s's realization. But the biconditional is false. Because causal influence is asymmetric, it may be that o's realization influences s's realization whereas s's realization does not influence o's realization.

Rational choice, however, aims at efficacy. An option's utility, within decision theory, assesses its efficacy. In obtaining an option's utility from the utilities of chances for possible outcomes, a relevant consideration is the option's causal influence on a possible outcome's probability, but not the evidence that the option provides for the possible outcome. Actions have utility because they promise to make good things happen, not because they produce evidence that good things will happen.

To obtain a version of expected utility analysis that accurately evaluates an option's utility, one must put aside evidentially orientated conditional probabilities. They serve well in statistics and inductive reasoning. But for expected utility analysis, which has a pragmatic motivation, one needs a causally directed type of conditional probability. So I dismiss standard conditional probability as an interpretation of $P(s$ given $o)$. That interpretation upsets the calculation of utilities of chances. It skews the probability weights for possible outcomes so that they do not reliably yield the utilities of chances for possible outcomes.[11]

An attractive, alternative interpretation of $P(s$ given $o)$, the probability of s if o were realized, takes the quantity as equal to the probability of the conditional that s would obtain if o were realized, which I abbreviate as $P((s$ if $o))$. I put in extra parentheses to emphasize that the probability applies nonconditionally to the entire conditional. The probability of the conditional, given a suitable reading in the style of Stalnaker (1981a), is directed by the causal influence of the option on the state rather than by the evidence the option provides for the state. The conditional $(s$ if $o)$ is true, roughly, if s obtains in the metaphysically possible world reached by holding fixed events prior to o, adding o to them, and then as much as possible letting causal laws generate o's aftermath. In particular, the supposition of o does not involve supposing a past altered to accommodate o. The forward direction of causation prohibits such backtracking. Notice that the conditional $(s$ if $o)$ may be true even if o does not cause s. Under the supposition that o, s may obtain either independently of o or because o causes s. The

---

[11] The principle of pros and cons does not rule out inclusion of irrelevant considerations, but if included, they must receive zero weight. Expected utility analysis is not upset by including evidential considerations, but by assigning them nonzero weight. Use of standard conditional probability assigns nonzero weight to evidential considerations in the calculation of the utility of a chance for a possible outcome.

conditional's truth is sensitive to causal relations but does not rely on them.[12]

The causal, nonbacktracking interpretation of the conditional has the following important consequence. If the state is believed to be causally independent of the option, the probability that the state would obtain if the option were realized is the same as the probability that the state obtains, whatever the option's evidential import concerning the state. That is, in this case $P((s \text{ if } o)) = P(s)$ even if $P(s/o) \neq P(s)$. Hence, taking $P(s \text{ given } o)$ as $P((s \text{ if } o))$ in Newcomb's problem, the probability of a state given an option is the same whatever the option.

The type of supposition of o in $P((s \text{ if } o))$ agrees with the type of supposition of o in $U(o)$ and $U(O[o \text{ given } s])$. Both the probability and the utilities consider what would happen if o were realized. Hence, $P((s \text{ if } o))$ equals the probability of the possible outcome $O[o \text{ given } s]$ if o were realized. Taking $P(s \text{ given } o)$ as equal to $P((s \text{ if } o))$ therefore suits expected utility analysis's pros and cons justification. It allows the conditional probability of a state to substitute for the conditional probability of a possible outcome in the pros and cons analysis of an option's utility, and so lets the principle of pros and cons support expected utility analysis.

The conflict between expected utility and dominance can be resolved if one maximizes expected utility using probabilities of subjunctive conditionals or some equivalent procedure. Because the choice has no causal influence on the prediction, the probability that if one-boxing were chosen it would have been predicted is just the probability that one-boxing has been predicted, and similarly for two-boxing. Thus the probabilities used to compute expected utilities are the same for each option, and the extra gain from two-boxing is decisive. It makes two-boxing's expected utility the greater. Calculating an option's expected utility using $P((s \text{ if } o))$ instead of ordinary conditional probabilities makes the principle to maximize expected utility more accurate. Given this revision, it correctly handles Newcomb's problem and other similar problems.

Gibbard and Harper (1981) adopt this resolution of Newcomb's problem. It relies heavily on causal concepts, which fill out the interpretation of the subjunctive conditionals involved. When causal

---

[12] The interpretation of conditionals is a subtle and controversial topic. I sketch only details relevant to my points about expected utility analysis. For an introduction to the literature, see Jackson (1991).

concepts are used to formulate the rule to maximize expected utility, the resulting decision theory is called a *causal* decision theory in contrast with an *evidential* decision theory that formulates the rule using ordinary conditional probabilities. Gibbard and Harper's causal decision theory recognizes that rational choice is a matter of efficacy rather than auspiciousness.

Contextualism condones causal concepts and welcomes causal decision theory. In contrast, operationism rejects causal decision theory because the concept of causality is unacceptable from its point of view.[13] Operationism opposes causal concepts in decision theory no matter whether they are introduced directly, as by Lewis (1981) and Skyrms (1980: pt. IIC), or indirectly through subjunctive conditionals, as by Stalnaker (1981b) and Gibbard and Harper (1981). Despite difficulties dealing with Newcomb's problem, operationists are inclined to accept evidential decision theory because of its conceptual parsimony. One understands the operationist's desire for conceptual economy, but truth has higher priority and causal concepts are needed for an accurate theory.[14]

Contextualism authorizes causal decision theory, and I endorse its resolution of Newcomb's problem. However, Gibbard and Harper's formula for expected utility needs some refinements to broaden its scope. A technical problem arises if $P(s$ given $o)$ is taken as equal to $P((s$ if $o))$. Sometimes the truth value of a subjunctive conditional is indeterminate. So for some option o and state s it may be indeterminate whether if o then s. Then the expected utility of o cannot be computed using the probability of the conditional that if o then s. Because the conditional is indeterminate, it does not have a probability. For example, consider betting that heads will come up on a coin toss. It may be indeterminate whether, if one were to bet, heads would come up. Maybe it would and maybe it would not. Then to calculate the expected utility of the bet, one cannot use the probability that if one were to bet, heads would come up. There is no such probability. Yet the probability

---

[13] The only way to make operationist sense of causality is to explicate it in terms of regularity conditions in the style of Hume. Carnap (1966: pt. IV) takes this approach. But regularity explications provide substitutes for causality, not the genuine article.

[14] See Sobel (1994: chap. 2) on the shortcomings of evidentialist attempts to resolve Newcomb's problem. Sobel observes that Eells's (1982: chap. 8) method for making evidential decision theory yield two-boxing faces objections in other decision problems, and in Newcomb's Problem supports the right choice for the wrong reasons.

that heads would come up given that one were to bet exists. Because $P((s$ if $o))$ does not exist in every case where $P(s$ given $o)$, intuitively taken, exists, $P(s$ given $o)$ cannot be taken as $P((s$ if $o))$.

Strictly speaking, the idealization about the availability of the probabilities needed for expected utility analysis puts aside the problem raised in the preceding paragraph. According to the idealization, $P((s$ if $o))$ exists if that probability is required by an expected utility analysis. Nonetheless, idealizations should be minimally restrictive. So expected utility analysis should be formulated to make the idealization about the existence of required probabilities as unrestrictive as possible.

To provide for the existence of $P(s$ given $o)$ in cases where $P((s$ if $o))$ does not exist, one may adjust the interpretation of the conditional $(s$ if $o)$ so that it is never indeterminate. One may adopt an interpretation of the conditional that dispels indeterminacy. According to Stalnaker's (1981a) analysis of conditionals, a conditional is true if and only if its consequent is true in the nearest antecedent-world. Adopting this interpretation, one may dispel indeterminacy by stipulating a unique nearest antecedent-world for every conditional arising in expected utility analysis.[15] This stipulation ensures that $P((s$ if $o))$ exists and equals $P(s$ given $o)$ in every case where the latter exists.

However, rather than provide an interpretation of the conditional $(s$ if $o)$ to make $P((s$ if $o))$ identical to $P(s$ given $o)$ in every case where the latter exists, it is more appealing from a theoretical perspective to abandon the attempt to reduce $P(s$ given $o)$ to the probability of a conditional, and instead take $P(s$ given $o)$ as a primitive. This tack saves applications of expected utility analysis from reliance on an unintuitive specification of nearest antecedent-worlds, one hard to complete and difficult to follow. So I abandon the attempt to find a quantity equal to $P(s$ given $o)$ in all cases, and to conclude the introduction of $P(s$ given $o)$ settle for a few further observations about $P(s$ given $o)$'s relation to other similar quantities.[16]

[15] The indeterminacy of conditionals is the reason the law of conditional excluded middle does not hold for conditionals as ordinarily understood. The indeterminacy is not dispelled, however, if the law of conditional excluded middle is made to hold by adopting the method of superevaluations as in Stalnaker (1981c). Under the method of superevaluations, some conditionals are still indeterminate.

[16] Some theorists try to deal with the influence of options on states by restricting the states used to compute expected utilities. Skyrms (1980: pt. IIC) restricts them to causal hypotheses about the consequences of options, and Lewis (1981) restricts

To clarify $P(s$ given $o)$, I also compare it to a kind of conditional probability introduced by Lewis (1976: 310), a *probability image*. The probability image of a state on an option is obtained by first transferring the probability of a world where the option is not realized to the nearest world where it is realized, and then adding the probabilities of the worlds in which the state obtains. For example, suppose that a Democrat, a Republican, and an Independent are the candidates in an election. One and only one will win. Each has the same probability of winning, namely, one-third. The Independent considers, only for the sake of argument, abandoning the race. Let d be the proposition that the Democrat wins, r the proposition that the Republican wins, and i the proposition that the Independent wins. Also, let a be the proposition that the Independent abandons the race. To obtain the probability image of d on a, first transfer the probability of $\sim a$-worlds to a-worlds. Let us suppose that if the Independent were to withdraw, his supporters would back the Democrat and supporters of the Republican and the Democrat would be steadfast. Hence, under imaging on a, all the probability of $(i \& \sim a)$-worlds shifts to d-worlds, all the probability of $(r \& \sim a)$-worlds remains on r-worlds, and all the probability of $(d \& \sim a)$-worlds remains on d-worlds. Next, add the probabilities of d-worlds after imaging on a. Because they collect all the probability of the i-worlds, the value of the probability image of d on a is two-thirds.

Probability imaging assumes that for each option and world there is a unique nearest option-world. When this assumption is met, for any

them to dependency hypotheses about the consequences of options. Lewis shows that, in cases where the restrictions hold, these approaches are equivalent to Gibbard and Harper's (1981) approach using probabilities of counterfactual conditionals – the approaches yield the same decisions. Despite their success handling causal relations, the restrictions on states are drawbacks from a theoretical point of view. They make expected utility analysis less general. Also, the restrictions are inexplicable given the pros and cons justification of expected utility analysis. One way to nullify these drawbacks is to specify a way of converting an expected utility analysis involving arbitrary states to an equivalent analysis involving causal hypotheses or dependency hypotheses. Note 22 explores this possibility for dependency hypotheses, and finds that it cannot be realized.

Skyrms claims that it is an advantage to use causal hypotheses in place of counterfactuals because counterfactuals are not well understood. Perhaps someone will claim that a restriction to causal hypotheses has a similar advantage over reliance on conditional probability taken as a primitive. This claim is related to the general operationist objection to contextualist decision theory and meets the same line of response. Operationist clarity has too high a cost in terms of theoretical desiderata such as generality and systematization.

option o and state s the conditional (s if o) is determinate and has a probability. The probability image is equal to the conditional's probability. Moreover, when the assumption is met, $P(s$ given $o)$, the probability of s if o were realized, equals the probability image of s on o.[17]

However, the probability image of s on o is not a fully adequate interpretation of $P(s$ given $o)$. $P(s$ given $o)$ exists in cases where the probability image of s on o does not exist. It does not presume a unique nearest o-world. $P(s$ given $o)$ equals the sum of the probabilities of the s-worlds after probability has been reassigned from ~o-worlds to o-worlds. But the reassignment does not assume that for every ~o-world, there is a nearest o-world. It takes place even if there are ties. All the probability of a ~o-world is reassigned to the nearest o-world where it exists, but in other cases the reassignment takes place in ways I do not attempt to specify. Thus $P(s$ given $o)$, being unreliant on a nearest o-world, exists in cases where $P(s$ if $o)$ and the probability image of s on o do not exist.

Because I do not specify the method of probability reassignment from ~o-worlds to o-worlds, my introduction of $P(s$ given $o)$ leaves gaps for intuition to fill. Further explication of $P(s$ given $o)$ is left for another time. Postponement is possible because our intuitive understanding of this primitive concept is firm enough to handle the work given it here.[18]

4.2.2. Utilities for the General Case

Dropping the assumption that options have no causal influence on states also requires a change in the utilities that figure in expected utility analysis. To make the case for change, I reconsider the utilities for option-state pairs used in Section 3.2. That section assumed that

---

[17] The interpretation of $P(s$ given $o)$ assumes constraints on the nearness relation that yields the probability image so that $P(s$ given $o)$ equals the probability image where it exists, just as the method of superevaluations assumes constraints on truth assignments so that the laws of logic are true with respect to each admissible truth assignment.

[18] For more on imaging, see Gärdenfors (1988: 108–18), Levi (1996: sec. 3.8), and Joyce (1999: 196–8, 257). Gärdenfors and Joyce generalize imaging for cases with no nearest o-world. They concentrate on the generalization's form and leave open substantive issues about the appropriate similarity relation between worlds. I suspend judgment on generalized imaging until more of the substantive issues are resolved.

they would be understood once utility and outcomes were clarified. I postponed a closer look at them until now. Their examination shows that they cannot accommodate an option's causal influence on states.

In Section 3.2 the utility for an option-state pair is $U(o$ given $s)$, the utility of o if s obtains. Suppose that o is taking vitamin C daily and s is the state that vitamin C prevents colds. Then to obtain $U(o$ given $s)$, follow a procedure akin to probability imaging. First transfer the probability of a world where vitamin C does not prevent colds to the nearest world where it does prevent colds. Then assess the utility of taking vitamin C daily with respect to worlds where vitamin C prevents colds. For someone skeptical about vitamin C's power to prevent colds, $U(o$ given $s) > U(o)$.

The example sketches a way to obtain $U(o$ given $s)$ in a case where a nearest s-world exists under an appropriate measure of distance between worlds. I offer no precise, general account of $U(o$ given $s)$ but take it as a primitive. It involves a basic psychological state of conditional desire. Section A.2 shows that for rational ideal agents the conditional utilities appearing in expected utility analysis have values derivable from unconditional utilities of worlds. But it does not treat conditional utility generally. Joyce (1999: chap. 7) presents a general representation theorem for conditional utilities, but to preserve conditional utility's explanatory power, I do not define it in terms of such representation theorems.

Although $U(o$ given $s)$ is a primitive, this section explains it by contrasting it with other similar quantities. The first point to emphasize is that $U(o$ given $s)$ is a conditional utility rather than the nonconditional utility $U(O[o$ given $s])$, although $U(o$ given $s) = U(O[o$ given $s])$ under the idealizations, as Section 3.2.3 states. Also, $U(o$ given $s)$, although conditional, is an actual utility assignment, not the hypothetical utility $U(o)$ given s. It meets accessibility requirements for the probabilities and utilities grounding expected utility analysis.

Given my interpretation of outcomes, and taking $U(o)$ as the utility of the outcome if o were realized, one may be tempted to claim that $U(o$ given $s)$ equals $U(o \, \& \, s)$. Both utilities seem to focus on the same possible worlds. $U(o \, \& \, s)$, being the utility of the outcome of $o \, \& \, s$, equals a probability-weighted average of the utilities of the possible worlds in which both o and s obtain. $U(o$ given $s)$'s value seems to be the same weighted average. According to this view of conditional utility, $U(o$ given $s) = U(o \, \& \, s) = U(s \, \& \, o) = U(s$ given $o)$. So $U(o$ given

135

s) = U(s given o); the outcome of o given s, and the outcome of s given o, are the same outcome, and so have the same utility. This consequence of the proposed equality plainly restricts it to types of utility that focus on outcomes rather than consequences. The commutativity of object and condition does not hold for a causally restricted type of utility, since causality is asymmetric. The equation of conditional utilities with utilities of conjunctions thwarts the goal of constructing a single method of expected utility analysis that applies alike to comprehensive and causal utility. This alone is sufficient reason to resist taking U(o given s) to equal U(o & s).[19]

Furthermore, even given the current comprehensive interpretation of utility, the equation of conditional utilities with utilities of conjunctions is wrong. Suppose one considers the utility of an option with respect to its negation. Is U(o given $\sim o$) equal to U(o & $\sim o$), as the general equation claims? U(o given $\sim o$) exists in some cases. It is the utility of the outcome if o were realized given that as a matter of fact o is not realized (the type of supposition for o varies with its role). On the other hand, U(o & $\sim o$) never exists. The logically impossible has no utility.[20] Because for some values of o and s, U(o given s) exists whereas U(o & s) does not, conditional utilities are not in general utilities of conjunctions.

This argument presumes the existence in some cases of the utility of an option given its negation. Does this utility exist in any case? Suppose I entertain the option of dining at Antoine's. If I know that I will not dine there, then U(d given $\sim d$) = U(d). If I know that I will dine there, the evaluation of U(d given $\sim d$) is more complex. To simplify, grant full knowledge of the actual world and suppose that there is a nearest $\sim d$-world w_1, and a d-world w_2 nearest w_1. Then to evaluate U(d given $\sim d$), I evaluate U(w_2). This utility may differ from U(d) if w_2 is not the actual world – if, for instance, under the supposition that I do not dine at Antoine's, the restaurant has a different, less skillful chef so that then if I were to dine there the meal would be less exquisite. An option's utility given its negation may be more difficult to evaluate in more complex cases, and may not exist in every case. But if it exists in some cases, then conditional utilities are not utilities of conjunctions.

---

[19] It is also sufficient reason to not to replace U(o given s) with U(s given o) or any other type of utility focused on s rather than o.

[20] Jeffrey (1983: 78) also puts necessary falsehoods out of preference rankings and out of utility assignments.

There is also a deeper reason why $U(o$ given $s) \neq U(o$ & $s)$ in some cases. To see it, first consider some implications of taking $U(o)$ as the utility of the outcome if o were realized. $U(o)$ involves a supposition about the option o. The supposition is expressed in the subjunctive mood, so I call it a *subjunctive* supposition. $U(o$ & $s)$ is similarly the utility of the outcome if o & s were realized. This quantity involves the subjunctive supposition of o & s, and so the subjunctive supposition of both o and s. In contrast, $U(o$ given $s)$ is the utility of the outcome if o were realized, given that s obtains. The supposition about the option o is expressed in the subjunctive mood, but the supposition about the state s is expressed in the indicative mood. I call it an *indicative* supposition.

Both types of supposition are relegated by rules sensitive to context. They may be analyzed in terms of nearest antecedent-worlds, but the nearness relations differ for each. In assessing nearness, subjunctive supposition gives priority to causal relations, whereas indicative supposition gives priority to evidential relations.[21] In typical contexts, subjunctive supposition of a condition keeps fixed events prior to the condition, adds the condition, and lets causal laws generate the aftermath. In typical contexts, indicative supposition of a condition modifies events surrounding the condition, both earlier and later, to reach the best evidential accommodation with the condition. This difference in the semantic import of indicative and subjunctive supposition may generate a difference in truth value for conditionals identical except for mood. For example, it is true that (1) if Oswald did not shoot Kennedy, someone else did. But it is false that (2) if Oswald had not shot Kennedy, someone else would have. The first conditional, being in the indicative mood, gives priority to evidential relations. Because our evidence is that Kennedy was shot, the supposition *if Oswald did not shoot Kennedy* generates a world where someone else did. The second condition, being in the subjunctive mood, gives priority to causal relations. Because, I assume, no one besides Oswald was motivated to shoot Kennedy, the supposition *if Oswald had not shot Kennedy* generates a world where no one shoots him.

---

[21] There are two methods of specifying the appropriate forms of supposition. One method is to identify features of types of supposition and indicate the features that the appropriate forms have. The other method is to identify types of supposition according to the linguistic factors, such as mood, used to express them, and use the identifying linguistic factors to indicate the appropriate forms of supposition. I take the latter approach. It is simpler because it relies on familiar linguistic distinctions. However, the former method is deepergoing.

To make expected utility analysis overcome Newcomb's problem, one needs subjunctive supposition of o and indicative supposition of s in $U(o$ given $s)$. Because in $U(o$ given $s)$ supposition of s is indicative whereas in $U(o \ \& \ s)$ supposition of s is subjunctive, it is not in general true that $U(o$ given $s) = U(o \ \& \ s)$. I therefore reject the equation of conditional utilities with utilities of conjunctions.

Having reviewed the interpretation of $U(o$ given $s)$, consider now its adequacy in a general formulation of expected utility analysis. Must one change the utilities used in expected utility analysis when states are not independent of options? If an option has some causal influence on the states used to compute its expected utility, one must, of course, make sure that the attractive or unattractive effects of this influence are counted in the addition of the utilities of the chances for the possible outcomes. Multiplying $U(o$ given $s)$ by $P(s$ given $o)$ does not suffice for this. Although the probability takes account of the influence of o on s, the utility must be modified to take account of the influence of o on s. One needs a causally responsive replacement for $U(o$ given $s)$.

To bring out the need for a replacement, consider the following case. John sees a notorious criminal at the airport. The criminal is about to escape to a foreign country. He will succeed unless John notifies the police, or unless someone else has already notified the police. John knows that there is a large reward for information leading to the criminal's arrest. And he knows that it is unlikely that anyone has already called the police. So the rational thing for him to do is to call the police. Unfortunately, inertia has a grip on John. He will not call and he knows this.

To simplify calculations, this case suspends the usual idealizations for expected utility analysis, which include rational decision making (see Section 3.2.5). The idealization of rational decision making puts aside cases where irrational decision making prompts compensatory departures from expected utility analysis. In my example I assume that John's irrational decision does not justify a violation of the expected utility principle. I also assume that the other idealizations are met, in particular, the input for an expected utility analysis of his options is flawless. The beliefs and desires involved in the analysis are fully rational. Under these assumptions, expected utility analysis generates the utilities of options. Moreover, deciding to call the police, although not an option he exercises, is still an option in the relevant sense. He can make that decision. Knowing that he will not make that decision does not remove his ability. So despite John's irrationality, the example bears on the formulation of expected utility analysis. Moreover, formulating expected utility analysis to handle it provides for the analysis's extension to nonideal cases.

Applying expected utility analysis using the utility of calling, given that the criminal will be apprehended, $U(c$ given $a)$, and other conditional utilities of this form, yields the following results. If the criminal will be apprehended, then since John will not report him, someone else has reported him already. So even if John were to call, he would not receive the reward. Therefore, because calling takes effort, $U(c$ given $a)$ is smaller than $U(\sim c$ given $a)$. I suppose that $P(a$ given $c) = 1$, and that $U(\sim c$ given $a) = U(\sim c$ given $\sim a)$ because, whatever the criminal's fate, John receives no reward if he does not call. Then the expected utility of calling is lower than the expected utility of not calling. Expected utility comparison yields the wrong decision.

The error arises because of a miscalculation of the utility of the chance for a reward. $U(c$ given $a)$ ignores calling's influence on apprehension. Because it assumes apprehension, it puts aside the possibility that calling causes apprehension. In general, suppose that it is desired that o cause s. This desire does not boost $U(o$ given $s)$, even if it is believed that o causes s. The indicative supposition that s makes the belief idle. The supposition makes s part of the background for entertaining o, and so precludes o's causing s. Although $U(o$ given $s)$ is comprehensive and entertains worlds, not just causal consequences, the supposition of s carries implications about causal relations and so directs evaluation to a world or set of worlds where o does not cause s. The conditional utilities used in expected utility analysis must have suppositions that direct evaluation to the right world or set of worlds.

To obtain a utility that registers the influence of o on s, instead of conditionalizing on s I conditionalize on the conditional that s if o. I replace $U(o$ given $s)$ with $U(o$ given $(s$ if $o))$. The latter quantity is the utility of the outcome if the option *were* realized, given that it *is* the case that the state *would* obtain if the option *were* realized. In the latter quantity s is supposed subjunctively as part of a condition supposed indicatively. The change in type of supposition for s makes the utility sensitive to o's causal influence on s. The utility is higher if it is believed that o causes s because the supposition $(s$ if $o)$ leaves it open that o causes s. It is unlike the supposition that s obtains, with its implication in the context of $U(o$ given $s)$ that s obtains independently of the option realized. Using the subjunctive conditional as the condition for an option's utility accommodates cases in which the option has a desirable or undesirable influence on the state. It makes expected utility analysis count that relevant consideration.

In my example, applying expected utility analysis using $U(c$ given $(a$ if $c))$, and other such "doubly conditional" utilities, yields the right decision. $U(c$ given $(a$ if $c))$ is much larger than $U(\sim c$ given $(a$ if $\sim c))$ because the first utility does not involve the supposition that John's calling is superfluous. It allows for the influence of John's belief that if he calls, he will be responsible for the criminal's apprehension.[22]

Next, consider another example showing how doubly conditional utilities capture relevant considerations otherwise lost. Its expected utility analyses have a few more steps than the previous example's but operate under all the usual idealizations. Imagine the final seconds of a basketball game. One team is ahead by a point. The other team shoots and misses. The ball is loose on the floor, about to roll into the hands of Reggie, a player on the team trailing. A player on the team leading, Kobe, wonders whether to dive to the floor to grab the loose ball. If he gains possession, time will run out and his team will win because of his

---

[22] Using $U(o$ given $(s$ if $o))$ as the utility for an option-state pair is roughly the same as requiring states to be dependency hypotheses, as in Lewis's (1981: 11–14) version of causal decision theory. Under the restriction to dependency hypotheses, the states for expected utility analysis have the form "s if o." They are hypotheses about the effect of an option on the state of the world and are themselves causally independent of options.

To compare methods of analysis, I first provisionally assume that dependency hypotheses have determinate truth values as a result of the existence of a unique nearest o-world. Then given their causal independence of options, one may simplify the application of my form of expected utility analysis with respect to a partition of dependency hypotheses. One may replace $P([s$ if $o]$ given $o)$ with $P((s$ if $o))$, which equals $P(s$ given $o)$ given a unique nearest o-world. Also, one may replace $U(o$ given $([s$ if $o]$ if $o))$ with $U(o$ given $(s$ if $o))$. These replacements allow derivation of my version of expected utility analysis from its application with respect to a partition of dependency hypotheses.

Now I discharge the assumption of nearest o-worlds. The unrestricted form of expected utility analysis no longer follows from its restriction to partitions of dependency hypotheses. The unrestricted form does not sacrifice accuracy but gains generality. For a set of dependency hypotheses of the form "s if o" is not necessarily exhaustive. If nearest option-worlds are not unique, the dependency hypotheses might all be indeterminate. Then the restricted form of expected utility analysis does not apply. The unrestricted form of expected utility analysis, however, still applies. Although $P((s$ if $o))$, a quantity appearing in a restricted analysis, requires a determinate truth value for the conditional $(s$ if $o)$, $U(o$ given $(s$ if $o))$, a quantity appearing in an unrestricted analysis, does not require a determinate truth value for that conditional. Whereas probability may not be attributed to a conditional of indeterminate truth value, utility may be assessed with respect to a condition of indeterminate truth value.

140

heroics. On the other hand, if Reggie gets the ball, he will take a final shot. If it goes through the hoop, Reggie's team will win. Kobe's diving after the ball involves some pain and risk of injury. Moreover, even if he doesn't dive, his team may win anyway. After getting the ball, Reggie may miss his shot. Should Kobe dive for the ball?

Suppose we compute the utility of Kobe's diving for the ball using simple conditional utilities. Taking as states (1) that his team will win and (2) that his team will not win, the formula states that $U(d) = P(w$ given $d)U(d$ given $w) + P(\sim w$ given $d)U(d$ given $\sim w)$. Consider $U(d$ given $w)$, the utility of Kobe's diving for the ball given that his team will win. Under the supposition that Kobe's team will win, his diving for the ball is superfluous – if Reggie were to scoop up the rebound and shoot, he would miss. Diving's superfluousness makes $U(d$ given $w)$ low. Thus the analysis may conclude that $U(d)$ is less than $U(\sim d)$.

The foregoing calculation of $U(d)$ ignores a relevant consideration, however. Kobe's team may win either because he dives and grabs the rebound, or because, although Reggie gets the rebound, he misses his shot. In the first case, Kobe's diving for the ball is the reason his team wins. $U(d$ given $w)$ ignores this possibility. To account for it, I switch to $U(d$ given $(w$ if $d))$. This quantity depends on two ways in which it may be true that if Kobe were to dive for the ball his team would win. According to the first, he gets the rebound. His diving causes his team to win. He is a hero. This is the possibility $U(d$ given $w)$ ignores. According to the second, he does not get the rebound. Reggie gains possession but misses his shot. Kobe's diving is superfluous. This is the possibility $U(d$ given $w)$ counts. The probability and utility of the first possibility may make $U(d$ given $(w$ if $d))$ large enough so that, using the formula $U(d) = P(w$ given $d)U(d$ given $(w$ if $d)) + P(\sim w$ given $d)U(d$ given $(\sim w$ if $d))$, an analysis may conclude that $U(d)$ is greater than $U(\sim d)$.

The trees in Figure 4.4 show more precisely that switching from simple conditional utilities to doubly conditional utilities may alter an option's utility ranking. They fill out the example by specifying probabilities and utilities. The descriptions at terminal nodes summarize considerations affecting utilities. The analyses assume that at each terminal node, except the one for Kobe's diving and winning, the simple conditional utility equals the doubly conditional utility. The bottom entries for $U(d)$ and for the terminal node for the pair d and w show the effect on $U(d)$ of switching to the doubly conditional utility $U(d$ given $(w$ if $d))$ at the terminal node. The tree for $U(d$ given $(w$ if $d))$ shows how

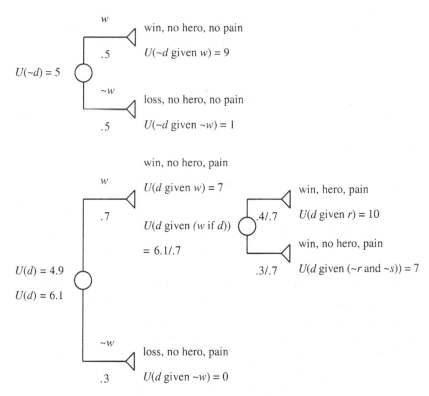

Figure 4.4. Switching to Doubly Conditional Utilities

it derives from the utility of Kobe's diving for the ball, given that he gets the rebound, $U(d$ given $r)$, and the utility of his diving for the ball, given that he does not get the rebound but Reggie does not score, $U(d$ given $(\sim r$ & $\sim s))$. It assumes that $P(r$ given $d)$ is .4 and that $P((\sim r$ & $\sim s)$ given $d)$ is .3. The latter is the product of Reggie's 60 percent chance of getting the ball despite Kobe's efforts and Reggie's 50 percent chance of making a shot.

One way of making do with simple conditional utilities is to use finer-grained states. For $U(d)$, one might use {Kobe gets the rebound, Kobe does not get the rebound and Reggie scores, Kobe does not get the rebound and Reggie does not score}. The first and third states make explicit the two possibilities influencing $U(d$ given $(w$ if $d))$. The simple conditional utility the first state generates, $U(d$ given $r)$, brings into $U(d)$'s calculation the possibility $U(d$ given $w)$ omits. Figure 4.5 shows

142

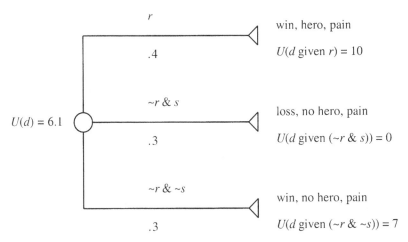

r

win, hero, pain

$U(d$ given $r) = 10$

.4

$U(d) = 6.1$

~r & s

loss, no hero, pain

$U(d$ given (~r & s)) = 0

.3

~r & ~s

win, no hero, pain

$U(d$ given (~r & ~s)) = 7

.3

Figure 4.5. Switching to Finer-Grained States

the results. However, retaining simple conditional utilities by restricting expected utility analysis to fine-grained partitions of states is very costly. Practicality demands preserving expected utility analysis's generality. Using doubly conditional utilities does this.

Another strategy for saving the simpler form of expected utility analysis is to adjust the form of supposition for a state in $U(o$ given $s)$ to suspend the assumption that s obtains independently of o. However, utility analysis is easier to apply if it relies on normal, intuitive types of supposition and its structure adjusts to accommodate an option's influence on states. Using $U(o$ given (s if o)) does the trick. It serves important practical objectives of expected utility analysis.[23]

To summarize, expected utility analysis involves suppositions about options and states. To make it precise, one must specify the sorts of supposition involved. I take the utility of an option to involve subjunctive supposition of the option. It evaluates the outcome if the option were performed. Subjunctive supposition (under its normal interpretation in decision problems) is directed by causal relations. Specifying subjunctive supposition of the option yields

---

[23] The remarks on future contingencies in Weirich (1980), discussed by Davis (1982), are intended to show that one cannot avoid the problem with $U(o$ given $s)$ simply by adjusting the form of supposition for s. Although this article addresses a type of causal utility, the point applies to comprehensive utility also.

143

an accurate version of the rule to maximize utility given the idealizations. The main alternative, taking the utility of an option to involve an indicative supposition of the option, yields an incorrect version of the rule. For indicative supposition (under its normal interpretation in decision problems) is directed by evidential relations. Hence the outcome if an option is performed may be attractive because of the evidence the option provides rather than because of its promise of good effects. Maximizing evidential attractiveness leads to bad decisions in cases such as Newcomb's problem.

Given that subjunctive supposition of an option is adopted for $U(o)$, expected utility analysis of that quantity also needs subjunctive supposition of the option for $U(o$ given $s)$. Theoretical simplicity then recommends subjunctive supposition for the option in $P(s$ given $o)$. In contrast, the supposition of the condition in $U(o$ given $s)$ should be indicative since the condition is taken as background for the option. Supposition of the condition should be directed by the evidence the condition provides rather than by the condition's effects. Also, strictly speaking, the condition should be the conditional that s if o. Although the conditional is supposed indicatively, its components are supposed subjunctively for the sake of uniform supposition of the option throughout expected utility analysis. Adopting these types of supposition for options and states yields an accurate form of expected utility analysis. According to it,

$$U(o) = \Sigma_i P(s_i \text{ given } o)U(o \text{ given } (s_i \text{ if } o)).$$

To handle a technical problem, the summation ranges over only s_i such that it is possible that s_i if o. This restriction is necessary because utility is not defined with respect to an impossible condition (recall earlier remarks about utility being undefined for an impossible proposition). Given my interpretation of the suppositions involved, the restriction ignores only s_i for which $P(s_i$ given $o)$ is zero. If utilities with respect to an impossible condition were given a default value, and included in the summation, they would not alter the value of the sum. Also, notice that there are no restrictions on the partition of states. The restriction is on the summation only. When applying expected utility analysis, I usually take the restriction as understood and do not restate it.

My formulation of expected utility analysis is general; it does not assume that states are independent of options. Under the idealizations, utility assignments comply with it. The principle of pros and cons supports the formula for an option's utility. The doubly conditional utilities prevent omission of relevant causal considerations.

Although the general formulation of expected utility analysis needs doubly conditional utilities, routine applications may reduce them to simpler utilities. In many cases, the agent's indifference to causal subtleties simplifies the formula's application. For example, suppose that someone with a headache is deciding whether to take an aspirin. To obtain the utility of taking an aspirin, she uses as states that her headache continues and that it does not. Applying the formula yields the equation, $U(a) = P(h$ given $a)U(a$ given $(h$ if $a)) + P(\sim h$ given $a)U(a$ given $(\sim h$ if $a))$. In evaluating $U(a$ given $(h$ if $a))$, the agent considers taking an aspirin under the supposition that if she were to take an aspirin her headache would continue. The supposition presents the evidentially nearest world in which the subjunctive conditional is true. In that world taking an aspirin is not efficacious. If the only relevant factors are the headache and the bother of taking medication, $U(a$ given $(h$ if $a))$ equals $U(a$ given $h)$, which in turn equals $U(a \& h)$. In evaluating $U(a$ given $(\sim h$ if $a))$, the agent considers taking an aspirin under the supposition that if she were to take an aspirin her headache would stop. The supposition presents the evidentially nearest world in which the subjunctive conditional is true. In that world taking an aspirin is either efficacious or superfluous. If, however, the only relevant factors are the headache and the bother of taking medication, $U(a$ given $(\sim h$ if $a))$ equals $U(a$ given $\sim h)$, which in turn equals $U(a \& \sim h)$. Then if $P(\sim h$ given $a)$ is large enough to compensate for the bother of taking medication, taking an aspirin is rational. The doubly conditional utilities reduce to simpler utilities in this case because the agent's concerns are simple. Applying my formula for expected utility reduces to applying other simpler formulas in such routine cases.

Section A.2 confirms my formula for an option's utility. It shows that expected utility analysis using doubly conditional utilities reduces to a fine-grained expected utility analysis using worlds as states and unconditional utilities of worlds. That is, an option's utility calculated using doubly conditional utilities agrees with its utility calculated using the formula $U(o) = \Sigma_i P(w_i$ given $o)U(w_i)$. The demonstration relies on the equation $U(o$ given $(w_i$ if $o)) = U(w_i)$. This equation holds because utility evaluates outcomes and because a possible world completely specifies an option's outcome, including all relevant causal factors.

The formula using worlds is accurate but requires taking states as worlds. My formula is more general. It does not require that states be worlds. It provides for coarse-grained analyses of utility in cases where causal relations between options and states matter. The price of this generality appears in my formula's utilities. They are doubly conditional utilities. The complication is a trifle, however, compared with the theoretical and practical advantages of a general form of expected utility analysis.

4.A. APPENDIX: ANALYSIS OF CONDITIONAL UTILITIES

Let us consider expected conditional utility. It arises if one expands a decision tree at a node for $U(o$ given $(s$ if $o))$, as one might if this quantity is not given in a decision problem but must be computed. In an expected conditional utility analysis of $U(o$ given $(s$ if $o))$, the condition $(s$ if $o)$ is broken down according to a set of states t_j forming a partition of states given s. I advance the formula

$$U(o \text{ given } (s \text{ if } o)) = \Sigma_j P_o(t_j \text{ given } s)U(o \text{ given } ((s \ \& \ t_j) \text{ if } o)),$$

subject to the following gloss on the conditional probabilities employed.

P_o is a probability assignment conditional on subjunctive supposition that o. Supposition of o is subjunctive here, just as in $P(s$ given $o)$ in an expected utility analysis of $U(o)$. On the other hand, because in $U(o$ given $(s$ if $o))$ the condition $(s$ if $o)$ is supposed indicatively, for parity of supposition of this condition and the conditions into which it is broken down, in $P_o(t_j$ given $s)$ the condition that s is supposed indicatively. The state s is supposed indicatively in $P_o(t_j$ given $s)$, even though it is supposed subjunctively within the condition $(s$ if $o)$ in $U(o$ given $(s$ if $o))$. The state s has a different role to play in the two contexts. In the utility it is taken as part of o's outcome, whereas in the probability it is used to determine o's outcome.

The indicative supposition of s in $P_o(t_j$ given $s)$ contrasts with the subjunctive supposition of o in $P(s$ given $o)$ in an expected utility analysis of $U(o)$. There the condition that o is subjunctively supposed to match o's subjunctive supposition in $U(o)$. It also contrasts with the subjunctive supposition of o in $P_o(t_j$ given $s)$. There the condition that o is supposed subjunctively to match o's subjunctive supposition within the condition (s if o) in $U(o$ given (s if o)). In general, the appropriate kind of supposition for a proposition depends on its role and the context. In my conditional probabilities, conditions that are options are subjunctively supposed, whereas conditions that are states are indicatively supposed.

I adhere to the same overall interpretation of conditional probability for expected conditional and nonconditional utility. But the interpretation produces a context-sensitive type of conditional probability. Whether a condition is an option or a state affects the way the condition is supposed, the way distance between worlds is measured, and the way probability images and conditional probabilities are determined. Given my interpretation of $P_o(t_j$ given $s)$, this quantity agrees with the P_o-image of t_j on s. Probability imaging on a condition just follows different nearness relations according as the supposition of the condition is indicative or subjunctive.

To support the displayed formula for $U(o$ given (s if o)), I also stipulate that states are indicatively supposed a special way when taken as conditions in the conditional probabilities used to obtain expected conditional utilities. To obtain a suitable separation of pros and cons, $P_o(t_j$ given $s)$ must equal $P_o(t_j$ & $s)/P_o(s)$. This equality's importance is apparent in cases where the option's possible outcomes given s have the same utility. Then the conditional probabilities $P_o(t_j$ given $s)$ must sum to one for the displayed formula to yield $U(o$ given (s if o)). The equality of $P_o(t_j$ given $s)$ and $P_o(t_j$ & $s)/P_o(s)$ ensures this result. To provide for this equality, one must adopt an appropriate kind of indicative supposition for s in $P_o(t_j$ given $s)$.

The kind of indicative supposition required is revealed by considering the case in which given o there is a unique nearest s-world. Then the indicative conditional $s > t_j$, taken as a nearness conditional, has a determinate truth value and so a probability under the idealizations. According to Lewis (1976: 311), $P_o(t_j$ given $s)$, given agreement with the P_o-image of t_j on s, is equal to $P_o(s > t_j)$, provided that the conditional $s > t_j$ uses the same nearness relation as imaging. But by Lewis's triviality theorem (1976: 300; 1986a), it is false that $P_o(s > t_j) = P_o(t_j$ & $s)/P_o(s)$ for nearness relations that are insensitive to the antecedents of conditionals. To obtain the identity $P_o(s > t_j) = P_o(t_j$ & $s)/P_o(s)$, as van Fraassen (1976) points out, one must choose a nearness relation that yields the independence condition

147

$P_o(s > t_j) = P_o(s > t_j/s)$. So I specify that $P_o(t_j$ given $s)$ incorporates a nearness relation that yields the independence condition.[24] Such a nearness relation is sensitive to context. It establishes distances between worlds in a way that depends on the antecedent of a conditional to which it applies. This makes the nearness relation radically context-sensitive in the eyes of some theorists, but the context sensitivity serves expected conditional utility analysis well.

[24] To obtain invariance of expected utility under expansion of trees, $P_o(t_j$ given $s)$ must equal $P_o(t_j \ \& \ s)/P_o(s)$, and so the independence condition must hold for indicative conditionals. See Weirich (1985). That article assumes invariance of expected utility analysis and derives the independence condition given probabilities of indicative conditionals as substitutes for conditional probabilities in the current formulation of expected conditional utility analysis. It treats the case in which states are independent of options. I stipulate that the interpretation of indicative conditionals required in this special case holds for the general case.

5

Two-Dimensional Utility Analysis

Analysis of the social utility of a new safety standard in Section 1.1 used two dimensions, people and their goals. Chapters 3 and 4 introduced a third dimension, possible outcomes. I have not yet presented group utility analysis, in which the dimension of analysis is people; that is reserved for Chapter 6. But having presented intrinsic utility analysis and expected utility analysis, in which the dimensions of analysis are goals and possible outcomes, respectively, I may undertake a two-dimensional analysis of an option's utility for a person.

A two-dimensional analysis expands a point in one dimension of analysis along the other dimension of analysis. In this chapter's two-dimensional analysis, the expansion takes place at an outcome, a point in the dimension for expected utility analysis. A possible world serves as the outcome, and its utility equals its intrinsic utility. Hence an intrinsic utility analysis may decompress its utility along the dimension of goals.

The two-dimensional framework makes my decision space more versatile. Using it, intrinsic and expected utility analyses may conjointly dissect an option's utility. This chapter explores the analytic fruits of two-dimensionality. It uses intrinsic utilities to compute comprehensive utilities of options given uncertainty.

5.1. INTRINSIC AND EXPECTED UTILITY ANALYSES COMBINED

Chapter 1 outlined multidimensional utility analysis. It involves the conjoint application of two or more independent forms of utility analysis. Intrinsic utility analysis and expected utility analysis are independent forms of utility analysis. Intrinsic utility analysis breaks down intrinsic utility according to basic intrinsic attitudes (BITs). Expected utility analysis breaks down comprehensive utility according to

149

possible outcomes or worlds. To apply these forms of analysis conjointly, a two-dimensional analysis needs a link between intrinsic and expected utility. The link is the identity of a world's intrinsic and comprehensive utility. Given this link the analysis may first break down an option's utility using worlds, and then break down a world's utility using BITs. For example, it may break down a safety policy's utility for a person using worlds that detail the policy's possible outcomes, and then break down each world's utility for the person using her basic intrinsic attitudes toward security, health, prosperity, and the like.

The two-dimensional framework for an option's utility analysis gives pros and cons finer grain than they have under a unidimensional analysis. With respect to the two-dimensional framework, a fundamental pro or con is a chance for a BIT's realization. Given the idealizations, no reasons for personal utilities have finer grain, and all other reasons are reducible to these. A world with a chance of realization is a package of chances for BITs' realizations. Various ways of repackaging the fundamental pros and cons yield derivative forms of utility analysis. For instance, an option's utilities given various worlds may be packaged as the option's utility given a state, so that the basic form of expected utility analysis using worlds yields the more general form using states.

Intrinsic and expected utility analyses use basic dimensions of analysis because the points at which they locate utilities (comprehensive or intrinsic) are sources of, as well as locations for, utility. Locating utilities at worlds creates a dimension of analysis because utilities so located may be added to obtain an option's utility. Similarly, locating utilities at BITs creates a dimension of analysis. But utilities at worlds and BITs are also sources of an option's utility. They explain its utility and do not merely provide a means of calculating it. Alternative forms of personal utility analysis, such as temporal utility analysis and causal utility analysis, are derivative. They do not introduce new explanatory factors. They do not introduce finer-grained and deeper reasons for or against an option. They merely provide other ways of locating utilities whose sum yields an option's utility.

5.2. INTRINSIC UTILITY ANALYSIS OF COMPREHENSIVE UTILITIES

This section presents new ways of using intrinsic utilities to analyze comprehensive utilities. These forms of intrinsic utility analysis assess basic reasons for an agent's extrinsic attitude toward a proposition.

Some analyses presented take account of information as well as basic intrinsic attitudes. One is the fundamental form of analysis to which others are reduced in the Appendix on consistency.

I call this chapter's forms of utility analysis intrinsic, although they analyze comprehensive utilities. The name "intrinsic utility analysis" comes from the intrinsic utilities that serve as the units of analysis. The analysans, not the analysandum, involves intrinsic utility. Intrinsic utility analysis dissects comprehensive utilities using intrinsic utilities, just as expected utility analysis dissects comprehensive utilities using expectations. It uses intrinsic utilities as building blocks, in the way that expected utility analysis uses expectations as building blocks. I apply the name "intrinsic utility analysis" whenever intrinsic utilities are units of analysis even when they are combined with probabilities to accommodate uncertainty.

This chapter's forms of intrinsic utility analysis assume that propositions are standardly named so that ideal agents are in a position to knowingly comply (see Section 3.2.3). They also assume a finite number of BITs and, consequently, a finite number of possible worlds, trimmed to be distinguishable only according to BITs realized. In virtue of this assumption a possible world may be represented by a finite set of BITs realized in the world, and a proposition may be represented by a lottery over a finite number of possible worlds. The assumption of a finite number of BITs is not burdensome, because, although I treat ideal agents, human agents are the ultimate targets. The assumption is not essential, in any case. My methods of intrinsic utility analysis, which separate reasons according to the principle of pros and cons, can accommodate an infinite number of BITs and worlds. The restriction to a finite number is for technical convenience. It simplifies the mathematics without essential loss of generality, and it tables issues such as the appropriate type of integral calculus and the appropriate principle of countable additivity for utility.[1]

5.2.1. Certainty

To begin, I assume certainty, for any proposition, of the possible world that would obtain if the proposition were true. Because the utility of a

---

[1] My method of utility assignment is similar to the method of probability assignment that attributes probabilities only to certain sets of points in a sample space with an infinite number of points. I assign utility only to certain sets of untrimmed possible worlds, those sets which represent trimmed worlds.

proposition evaluates the outcome if the proposition were true, if the proposition's world is certain, then the proposition's utility is the same as that world's utility.[2]

Now the utility of a possible world is not affected by uncertainty about what might or might not obtain if the possible world were realized. The possible world is a full specification of what would happen if it were realized; there is no uncertainty about the effects of its realization. In this way utilities of worlds are like intrinsic utilities of worlds. In fact, since every feature of a world is a logical consequence of the world (a maximal consistent proposition), there is no difference at all between the utility and intrinsic utility of a possible world, as Section 2.3 explained. The intrinsic utility of a possible world, taken with respect to its logical consequences, is the same as its utility, taken with respect to its total outcome, including causal consequences. Given certainty, we may therefore analyze $U(p)$ in terms of intrinsic utilities. $U(p)$ is the utility of p's world, which is the intrinsic utility of p's world, and this is a sum of the intrinsic utilities of the objects of the BITs realized in the world. Hence $U(p)$ is the sum of the intrinsic utilities of the objects of the BITs realized in p's world. Letting $W[p]$ stand for p's world, and letting $O[p]$ stand for the proposition that $W[p]$ obtains, the formula for $U(p)$ given certainty is:

$$U(p) = U(O[p]) = U(W[p]) = IU(W[p]) = \Sigma_j IU(\text{BIT}_j),$$

where BIT_j ranges over the objects of the BITs realized in $W[p]$.

The first of this string of identities holds in virtue of my interpretation of U. The second presumes certainty about p's world. For given uncertainty about p's world, the utility of the proposition that p's world obtains is the utility of a lottery over possible worlds where p holds – a probability-weighted average – whereas the utility of p's world is the utility of a particular world. The third identity holds in virtue of the identity of U and IU for worlds. The last identity holds in virtue of Section 2.2.2's summation principle for worlds.

Commodity bundle analysis in economics, given a standard set of assumptions for it, is a form of intrinsic utility analysis given certainty. Commodity bundles, dated to indicate time of possession, are taken as

---

[2] Cases in which none of the proposition-worlds emerges as the proposition's world may be treated as cases of uncertainty about the proposition's world. Each proposition-world receives a probability given the proposition's realization, as in the next section.

possible worlds trimmed of irrelevant details. By assumption, there are basic intrinsic desires for amounts of the commodities that then cause intrinsic desires for the commodity bundles. A standard assumption is the additive separability of intrinsic utilities of the amounts of the commodities in a bundle. Accordingly, the intrinsic utility of the bundle is an additive function of the intrinsic utilities of the amounts of the commodities. The intrinsic utility of a bundle is then taken to be its comprehensive utility. This form of intrinsic utility analysis appears in Keeney and Raiffa's (1976: chap. 3 and sec. 9.2) treatment of decisions with multiple objectives, freely interpreted.

5.2.2. Uncertainty

To generalize intrinsic utility analysis of comprehensive utilities, I now explore the connection between intrinsic and comprehensive utility given uncertainty. IU depends on logical consequences and so is independent of information, whereas U depends on information. To obtain U from IU given uncertainty, one must allow for the effect of information on U. Making the right adjustments, $U(p)$ can be obtained from intrinsic utilities of objects of BITs.

We do not always know the world that would be realized if a proposition p were realized. So an intrinsic utility analysis of $U(p)$ is not always as simple as calculating the intrinsic utility of p's world. Given uncertainty, the pros and cons that yield $U(p)$ are not reducible to the realization of a set of BITs. For it may be uncertain whether p's realization would lead to a particular BIT's realization. To handle this uncertainty, I treat a proposition as a lottery over possible worlds. I obtain utilities of chances for the worlds from the worlds' probabilities and intrinsic utilities, and from the utilities of the chances I obtain the utility of the proposition whose realization produces the chances. The rest of this section works out the details.

Belief affects utility because utility considers possible outcomes when outcomes are uncertain. Given uncertainty, I take a proposition's utility as the utility of a lottery over the possible worlds where the proposition is realized. The lottery's utility is its expected utility. The utility of a chance for a possible world is taken to be the probability of the possible world times its utility. As Section 4.1 explains, this expected utility analysis takes into account all factors affecting the proposition's utility, including factors such as a basic intrinsic aversion to risk. If the proposition's realization involves a risk, then worlds realizing the

153

proposition involve a risk. An aversion to risk lowers their utilities. For example, consider a gamble from Section 4.1's version of Allais's paradox, the gamble that pays $4,000 with a probability of 4/5 and $0 with a probability of 1/5. An agent who evaluates such gambles only with respect to return and risk may treat selecting the gamble and winning as a world in which he undertakes the gamble's risk and has a $4,000 return. If he is averse to risk, he discounts the world's utility because of the risk the world includes.

Next, I convert a proposition's expected utility analysis in terms of worlds into an intrinsic utility analysis. To do this, I replace the utilities of the worlds with their, equivalent, intrinsic utilities. When the tree for an expected utility analysis of a proposition is maximally expanded and has worlds at terminal nodes, the comprehensive utility of the proposition is the expected intrinsic utility of the proposition's world. That is, $U(p) = U(O[p]) = EU(W[p]) = EIU(W[p])$, where $O[p]$, the proposition that $W[p]$ obtains, is represented by a lottery over worlds for which p gives chances. Spelling out the expected intrinsic utility, I obtain the following form of intrinsic utility analysis:

$$U(p) = \Sigma_i P(w_i \text{ given p}) IU(w_i),$$

where w_i ranges over all possible worlds where p holds.

Section 2.2.2 shows that for a world w, $IU(w) = \Sigma_j IU(BIT_j)$, where BIT_j ranges over objects of BITs realized in w. Breaking down the intrinsic utility of worlds into the intrinsic utilities of objects of BITs realized in them, my form of intrinsic utility analysis yields this corollary:

$$U(p) = \Sigma_i P(w_i \text{ given } p) \Sigma_j IU(BIT_{ji}),$$

where w_i ranges over worlds where p holds and BIT_{ji} ranges over objects of BITs realized in w_i. The Appendix on consistency adopts this form of intrinsic utility analysis as the canonical, or foundational, form. It is fine-grained because it is a hybrid of expected and intrinsic utility analyses.

Intrinsic utility analysis differs from expected utility analysis concerning the conditionalization of utilities used as input. The formula for a proposition's expected utility needs conditional utilities because the proposition's realization may influence the states used to assess it. On the other hand, the formulas for a proposition's intrinsic utility analysis may use nonconditional intrinsic utilities. They need not conditionalize the intrinsic utilities of worlds or the objects of BITs because intrinsic attitudes toward these are independent of conditions, as Section 2.1.3 observes.

5.2.3. Chances for Realizations of BITs

Because the utility of a world derives from the intrinsic utilities of the objects of BITs realized in the world, $U(p)$ may also be analyzed in terms of the utilities of subjective chances for realizations of BITs. For each BIT, p's realization produces, as a sure, epistemic consequence, a subjective chance (dependent on a subjective probability) that the BIT is realized. Intrinsic utility analysis may take such chances as pros and cons and add their utilities to obtain $U(p)$. Because this form of analysis uses intrinsic utilities of objects of BITs, it is a form of intrinsic utility analysis, but it yields $U(p)$ because it also draws on information and probabilities.

This new form of intrinsic utility analysis obtains $U(p)$ more directly from BITs than the previous section does. It does not use the intrinsic utilities of worlds as intermediaries. It calculates $U(p)$ from probability-utility products for BITs. The calculation is not guided by expected utility analysis with respect to a partition of worlds. It uses a partition of BITs and probabilities of BITs' realizations rather than probabilities of worlds. The new form of analysis is a "BIT by BIT analysis," whereas the previous section's form is a "world by world analysis." It highlights the BIT dimension of analysis. The new form of analysis is, however, structurally similar to expected utility analysis. It treats realization of an agent's BITs as possible outcomes of p's realization. It assumes that there is an intrinsic utility of realizing each BIT, and a probability of realizing each BIT if p were realized. To obtain $U(p)$, it then takes the intrinsic utility of each BIT's object, weights it by its probability given p, and then adds the products. According to it,

$$U(p) = \Sigma_j P(\text{BIT}_j \text{ given } p)IU(\text{BIT}_j),$$

where BIT_j ranges over the objects of all BITs.

This analysis of $U(p)$ is a paradigm application of the principle of pros and cons. The considerations are chances for realizations of BITs. The nature of BITs ensures that there is no omission or double counting. The utility of a chance for the realization of a BIT is the probability of its realization times the intrinsic utility of its realization. The addition of the utilities of the chances yields the expected utility of p's outcome, which as Section 3.1 explains, equals p's utility.

A BIT by BIT intrinsic utility analysis of $U(p)$ in cases where p is a world w, a maximal consistent proposition, claims that $U(w) = \Sigma_j P(\text{BIT}_j$ given $w)IU(\text{BIT}_j)$, where BIT_j ranges over the objects of all BITs. $P(\text{BIT}_j$ given $w)$ is 1 if BIT_j's realization is entailed by w; otherwise it is 0. Hence the formula for $U(w)$ reduces to $U(w) = \Sigma_j IU(\text{BIT}_j)$, where BIT_j ranges over objects of BITs whose realization w entails. This is the same formula as obtained from a world by world intrinsic utility analysis of w. Using Section A.3's methods, one can show that an intrinsic utility analysis of $U(p)$ using probability-utility products for BITs is in general equivalent to an intrinsic utility analysis of $U(p)$ using worlds.

Because I have advanced two general methods of calculating an option's comprehensive utility, namely, expected utility analysis and intrinsic utility analysis, another question about consistency arises. Do the two methods yield the same results? Because both methods follow from the principle of pros and cons, they should yield the same results. Nonetheless, an independent verification of their consistency may confirm that the principle of pros and cons has been correctly applied in obtaining these methods of utility analysis. Section A.3 establishes their consistency by showing that both expected utility analysis and intrinsic utility analysis (in its BIT by BIT form) are derivable from Section 5.2.2's fundamental form of analysis involving probabilities of worlds and intrinsic utilities of objects of BITs.

5.3. PARTIAL UTILITY ANALYSIS

Section 5.2 presents two forms of intrinsic utility analysis of $U(p)$. They use intrinsic utilities of worlds or objects of basic intrinsic attitudes. They also use expected utility analysis to accommodate uncertainty and so are two-dimensional forms of analysis. This section presents a novel form of analysis suggested by the two-dimensional framework. It uses *partial utilities* defined in terms of restricted sets of BITs instead of intrinsic utilities of worlds or objects of BITs. The partial utilities, although restricted in consideration of BITs, are still comprehensive in consideration of outcomes.

A partial utility stems from desires and aversions an agent has putting aside certain considerations. These restricted attitudes are commonplace. For example, an agent might want to flavor foods with butter, putting aside concerns about cholesterol. When such restricted attitudes serve as the basis of utility, one may obtain the utility of a proposition with respect to some but not all considerations – say, the utility for the agent of using butter, while putting aside worries about cholesterol. A proposition's partial utility for an agent is the proposition's utility putting aside certain intrinsic desires and aversions of the agent. Assessments with respect to a partial set of BITs, especially assessment that puts aside all BITs save one, are very useful. I focus on the partial utility of a proposition with respect to a particular BIT. The partial utility evaluates the proposition as if that particular BIT were the only one. As I show here, partial utilities with respect to BITs may be added to obtain comprehensive utilities.

Partial utilities have a major theoretical role in some areas of philosophy. For example, to apply utilitarianism, Dworkin (1978: 234–8) uses desires for social policies, putting aside desires about other people's lots. Partial desires and aversions, the basis for partial utilities, also have an important role in descriptions of our mental life. Suppose someone is averse to going to the dentist, but wants to go all things considered, or on balance, because of the long-term benefits of dental care. The aversion to going to the dentist is not an aversion on balance; it ignores intrinsic attitudes such as an intrinsic desire for sound teeth. Also, the aversion is typically not an intrinsic aversion because going to the dentist does not entail any level of satisfaction stemming from realization of basic intrinsic desires or aversions. In a typical case the aversion to going to the dentist is a partial aversion that registers only some of the agent's intrinsic aversions, such as aversion to pain.

Many desires and aversions are partial rather than all things considered. The intrinsic attitudes they take into account are generally ones immediately associated with the object of desire or aversion. Some desires and aversions wear blinders, as do partial desires and aversions, but are distinct because of the type of blinder. For example, some desires and aversions put aside long-term consequences rather than BITs. Although a partial desire or aversion limits considerations, it limits the range of BITs considered, rather than the range of consequences considered.

In a partial utility assessment of a proposition, therefore, utility is not taken as intrinsic utility but as comprehensive utility. The assessment attends to the proposition's entire outcome including its realization's long-term causal consequences. Partial utility appeals to BITs, as does intrinsic utility, but unlike intrinsic utility uses them only to limit the criteria of assessment, not the scope of assessment. Also, unlike intrinsic utility, partial utility is sensitive to information. It handles uncertainty the way comprehensive utility does, through expected utility analysis. Partial utility equals expected partial utility.

Although partial utilities put aside desires, not consequences, partial utilities are equivalent to utilities that put aside consequences not relevant given the desires put aside. That is, they are equivalent to utilities of trimmed worlds. The utility of a trimmed world derives exclusively from the world's features, not also expectations concerning untrimmed worlds. It is the trimmed world's intrinsic utility computed with respect to BITs whose realization the trimmed world entails. As a result, partial utilities with respect to pleasure, say, are equivalent to utilities of worlds trimmed of everything but pleasure.

Generally, in applying expected utility analysis I use as a set of possible worlds, not full blown possible worlds, but trimmed possible worlds including all and only features that matter. These trimmed possible worlds are maximal consistent propositions with respect to BITs. Partial utilities, in effect, trim worlds even more. They trim not only irrelevant matters but even some relevant matters. Because a partial utility is a comprehensive utility with respect to trimmed worlds, given uncertainty it is equivalent to the expected utility of a lottery over worlds in which some features have been blacked out as in Figure 5.1.

To clarify partial utilities, I examine the relationship between the partial utility $U_j(p)$ and the intrinsic utility $IU(\mathrm{BIT}_j)$. $U_j(p)$ is the utility of p on the assumption that BIT_j is the only BIT. U_j is understood so that utility comes only from realization of BIT_j and not from nonrealization of BIT_j. Realization of BIT_j is the only state that matters to the partial utility. When it is certain that BIT_j would be realized if p were realized, $U_j(p) = IU(\mathrm{BIT}_j)$, because the only factor that makes $U(p)$ and $IU(\mathrm{BIT}_j)$ differ then is the existence of BITs besides BIT_j, and that factor is put aside by the partial utility $U_j(p)$. Also, when it is certain that BIT_j would not be realized if p were realized, $U_j(p)$ is zero, because from the standpoint of U_j there is intrinsic indifference to anything other than realization of BIT_j. When it is uncertain whether BIT_j would be realized if p were realized, $U_j(p)$ equals $IU(\mathrm{BIT}_j)$ times the probability that BIT_j would be realized if p were realized. That is,

$$U_j(p) = P(\mathrm{BIT}_j \text{ given } p)IU(\mathrm{BIT}_j).$$

This formula, because it comprehends the case of certainty, provides the canonical method of obtaining input for a partial utility analysis.

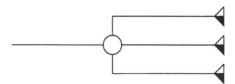

Figure 5.1. Partial Utility

The formula for $U_j(p)$ follows from an expected utility analysis using as a partition of states: realization of BIT_j and nonrealization of BIT_j. $U_j(p)$ is a probability-weighted average of the case in which BIT_j is realized and the case in which it is not. Hence $U_j(p) = P(BIT_j$ given $p)U_j(BIT_j) + P(\sim BIT_j$ given $p)U_j(\sim BIT_j)$. The utilities in the summands are simplified because with respect to U_j only realization of BIT_j matters. Given my interpretation of U_j, $U_j(BIT_j) = IU(BIT_j)$. Also, because from the perspective of U_j there is indifference toward the case in which BIT_j is not realized, the second product involves multiplication by zero and so drops away.

Now I formulate partial utility analysis using partial utilities. Because BITs separate relevant considerations so that there is no omission or double-counting, adding the partial utilities of a proposition with respect to each BIT yields the utility of the proposition. In other words,

$$U(p) = \Sigma_j U_j(p),$$

where U_j is utility with respect to BIT_j and BIT_j ranges over all BITs. In this formula the range of BITs may be reduced to those whose probabilities of realization would be positive if p were realized. Partial utilities with respect to other BITs are zero and do not affect the sum. The BITs with positive probabilities of realization given a proposition are paradigm pros and cons for an analysis of the proposition's utility. Intuitively, the weights of these pros and cons are the partial utilities of the proposition with respect to the pros and cons. Using BITs to separate considerations ensures that there is no omission or double-counting, and hence that the principle of pros and cons justifies adding the weights of the pros and cons to obtain the proposition's utility.

Notice that intrinsic utility analysis of a world's utility using intrinsic utilities of BITs' objects is equivalent to an intrinsic utility analysis of the world's utility using partial utilities. Take a BIT realized in the world. The intrinsic utility of realizing the BIT is the strength of the

159

BIT, positive for a basic intrinsic desire, and negative for a basic intrinsic aversion. It is the same as the partial utility of the world with respect to the BIT because the world's specification removes uncertainty about the BIT's realization. That is,

$$U_j(w) = IU(\text{BIT}_j),$$

if BIT_j is realized in w. When the utility of the world is obtained by adding the intrinsic utilities of the objects of BITs realized in it, the calculation is equivalent to adding the partial utilities of the world with respect to the BITs realized in it – and also with respect to all BITs because partial utilities of the world with respect to BITs not realized in it are zero.

Partial utility analysis handles uncertainty because the partial utility of a proposition with respect to a BIT responds to uncertainty that the BIT would be realized if the proposition were realized. It is an expected partial utility. One may make the treatment of uncertainty explicit by combining partial utility analysis with expected utility analysis. One may analyze partial utilities using probabilities as in expected utility analysis. The result is a hybrid form of analysis:

$$U(p) = \Sigma_j\Sigma_i P(s_i \text{ given } p)U_j(p \text{ given } (s_i \text{ if } p)),$$

where j ranges over BITs and i ranges over states in a partition. To make the treatment of uncertainty fully explicit, as opposed to implicit in the calculation of conditional partial utilities, one may use the partition of possible worlds as the partition of states. The partial utility of a possible world involves no uncertainty.

A conditional utility may also be analyzed using partial utilities. A simple version of the analysis states that $U(p \text{ given } t) = \Sigma_j U_j(p \text{ given } t)$, where j ranges over BITs. A version making the treatment of uncertainty explicit states that

$$U(p \text{ given } t) = \Sigma_j\Sigma_i P_p(s_i \text{ given } t)U_j(p \text{ given } ([s_i \& t] \text{ if } p)),$$

where j ranges over BITs and i ranges over states in a partition.

An analysis of $U(p)$ using utilities partial with respect to BITs is consistent with an analysis of $U(p)$ using intrinsic utilities of objects of BITs. This is expected given their derivations from the principle of pros and cons. Section A.4 demonstrates their consistency to confirm their derivations.

Intrinsic utility analysis is, of course, proposed under some idealizations. The idealizations for attainment of goals of rationality are the standard ones about the agent's cognitive abilities and rationality in matters related to desire formation. Under these idealizations, it is a goal of rationality to form desires so that the utility of an option equals its value computed by an intrinsic utility analysis, and agents are perfectly able to realize this goal. The other idealizations provide the wherewithal of intrinsic utility analysis. I assume that an agent has a set of BITs and that the realization of a BIT has a quantitative intrinsic utility. The idealizations are strong. Nonetheless, intrinsic utility analysis is useful for organizing deliberations in real-life decision problems.

The main objections to intrinsic utility analysis are operationist. Objectors observe that, unlike utilities, intrinsic utilities of objects of BITs and partial utilities with respect to BITs are generally not derivable from an agent's choices. An agent's choices generally do not reveal partial utilities, for instance, because choices rest on all relevant considerations, not just part of them. Only special choices or choices in special situations reveal partial utilities. For instance, the choices of a fair judge at an art show might indicate the paintings he likes best putting aside nonaesthetic factors such as his bias in favor of his nephew's paintings. They might indicate partial utilities with respect to the goal of fairness because they are made to promote fairness alone. Such revelatory choices are exceptional, however.[3]

Despite difficulties inferring partial utilities from choices and, consequently, difficulties defining partial utilities in a way that meets operationist standards, partial utilities are meaningful according to contextualist criteria of meaning. Although partial utilities have a limited role in an operationist decision theory, they have a major role in a contextualist decision theory. They have explanatory value justifying their inclusion.

Intrinsic utility analysis is very useful for explaining preferences. Preferences have a hierarchical structure. Some preferences are basic

---

[3] The crosscutting separability assumption, discussed, for instance, by Broome (1991: sec. 4.2), identifies cases in which choices reveal preferences with respect to a single dimension of utility assessment, which may be a single BIT. Also, Keeney and Raiffa's (1976: chap. 6) methods of multidimensional utility analysis, use overall preferences and choices, given the assumption of additive separability, to make inferences about utilities along a single dimension of utility assessment, which again may be a single BIT.

and others causally derivative. I want to explain preferences among worlds in terms of basic intrinsic desires, and I want to explain preferences among propositions in terms of beliefs and desires concerning worlds. Intrinsic utility analysis is well suited for this task, although intrinsic utilities of BITs' objects are difficult to identify in real-life cases. Encouraged by contextualism, I accept intrinsic utilities of BITs' objects as theoretical entities that explain an agent's choices in a deep-going way.

5.4. MEAN-RISK ANALYSIS OF UTILITIES

As an illustration of partial utility analysis of an option's utility, I examine *mean-risk analysis* of an option's utility. This form of utility analysis is used in portfolio theory. It is applied to financial decision problems where it is helpful to separate considerations for options into risks and other consequences, generally profit or return. It might be applied, for example, to a financial decision problem for a corporation with two basic intrinsic attitudes, desire for profit and aversion to risk. Given standard simplifying assumptions about the utility of money, the degree of risk involved in the corporation's ventures depends on its wealth, the extensiveness of the evidence on which its probability assignment rests, and the dispersion of probability-return products. For instance, the size of the risk involved in a venture that promises a fixed subjective chance for a fixed benefit is inversely proportional to the corporation's wealth.[4]

A common method of evaluating options known as *mean-variance analysis* is a form of mean-risk analysis. According to this method, an option is evaluated in terms of its expected, or mean, monetary return, and the variance of the probability distribution over the possible monetary returns. The variance is taken to indicate (approximately) the risk involved in the option. Where there is no variance, the option is a sure thing. Where there is a large variance, the option is risky. Given standard simplifying assumptions about the utility of money, mean-variance analysis claims that an option's utility is its mean return minus a proportion of the variance, the proportion depending on the degree to which the agent is averse to risk.

---

[4] As Brady (1993) points out, the precedents for combining expected value with risk assessments go back at least as far as John Maynard Keynes.

I mention mean-variance analysis only as an example of mean-risk analysis. There are many objections to mean-variance analysis.[5] Variance is not an accurate measure of risk. For instance, contrary to mean-variance analysis, variability in the domain of losses indicates more risk than variability in the domain of gains. Because of such objections, theorists seek measures of risk better than variance. I do not review the relevant literature because I do not propose a measure of risk. I just take the risk involved in an option as a contextually introduced theoretical entity, and consider the utility the agent assigns to it. The utility is not a measure of the risk but of the agent's attitude toward it. Tabling the measurement of risk is warranted, because my goal is only to sketch the justification of mean-risk analysis, which separates considerations into risk and other factors. The issue is whether this separation is warranted and not how to represent the terms of the separation, in particular, risk.

Given some assumptions about the agent's attitude toward risk, partial utility analysis justifies a form of mean-risk analysis. The main assumption is that the agent has a basic intrinsic aversion to risk. This assumption may be warranted in many cases because aversion to risk often seems to be intrinsic and not caused by other intrinsic aversions. The assumption is not warranted in all cases, however. Suppose that a person undertaking a risk feels butterflies in his stomach. Because the sensation is unpleasant, the aversion to butterflies may cause an aversion to risk. In this case the aversion to risk is not a basic intrinsic aversion.

Another assumption concerns the objects of aversion to risk. Because risks have various sizes, agents have aversions to particular risks, not just risk in general. Usually the larger the risk, the greater the aversion. I assume that the agent has a family of basic intrinsic aversions (BIVs), each being an aversion to risk of degree n at time t. Although the members of the family for time t have objects that are exclusive, I still assume that none causes another so that they qualify as BIVs. In evaluating an option, I take the relevant BIV to be a BIV to the particular risk involved in

[5] See, for example, Markowitz (1959: 193–4, 286–91, 296–7) and Raiffa (1968: 55–6). Some methods they discuss use the standard deviation, the square root of the variance, as opposed to the variance itself. Their objections carry over to methods using the variance.

the option rather than risk in general. For brevity, I do not spell this out in my example.[6]

Granted a basic intrinsic aversion to risk, one may use partial utility analysis to divide an option's utility into its utility with respect to aversion to risk and its utility with respect to other basic intrinsic attitudes. In symbols, $U(o) = U_r(o) + U_{r\sim}(o)$. Because expected utility analysis may be applied in conjunction with partial utility analysis, one may substitute an expected or mean utility for $U_{r\sim}(o)$. The result is a mean-risk analysis of $U(o)$. Partial and expected utility analyses together support the mean-risk utility analysis.

The foregoing justification of mean-risk utility analysis holds however risk is defined, provided that there is a basic intrinsic aversion to it. However, if risk is defined as in Section 4.1, as a subjective chance for gain or loss, then because risk depends on probabilities and utilities of outcomes (including risk), and because an ideal agent has unlimited cognitive power and full awareness of his mental states, an ideal agent is certain of the risk involved in an option. Hence $U_r(o)$ is available to him; it is a known quantity. Thus mean-risk utility analysis provides a useful method of calculating $U(o)$ given the idealizations.[7]

This appendix introduces additional forms of utility analysis. Some combine intrinsic and expected utility analyses in novel ways. One generalizes partial utility analysis.

5.A.1. Viable Hybrids

I distinguish the *consistency* of two methods of analyzing the same quantity, and the *compatibility* of two methods of analysis, such as

---

[6] Kahneman and Tversky (1979: 268–9) argue that people are attracted to taking chances in the face of possible losses. Having an option's utility assess the particular risk involved in the option allows for an agent's being averse to some risks, and indifferent or even attracted to others, although I put aside such cases for now.

[7] Temporal utility analysis also provides an intuitive justification of mean-risk analysis. Suppose an agent has a basic intrinsic aversion to risk (in the subjective sense). The agent incurs the risk involved in a risky option the moment he adopts the option. The realization of other BITs typically takes longer. A temporal approach ranks an option according to the BITs realized at the moment of its adoption and the BITs realized during its temporal sequel. When the temporal sequel is uncer-

expected utility analysis and intrinsic utility analysis, applied to different quantities. Independent methods of utility analysis may be incompatible in the sense that they are not accurate if applied simultaneously – say, one to an option and the other to an option-state pair, or one to a new type of utility introduced by the other. A *hybrid* form of utility analysis is a combination of two independent methods of utility analysis, different methods for different quantities or crossing types of utility. A *viable hybrid* is a compatible combination of independent methods of utility analysis.

To illustrate, take expected utility analysis applied to intrinsic utility. It is a viable hybrid. The expected value of $IU(p)$ equals $IU(p)$. Because intrinsic utility assesses only logical consequences and these are the same in all states, a probability-weighted average of $IU(p)$ with respect to a partition of states equals $IU(p)$. This hybrid analysis, although viable, is not very useful, however, because, in effect, it calculates $IU(p)$ from $IU(p)$. A more interesting hybrid of expected and intrinsic utility analysis is a world-by-world intrinsic utility analysis of $U(p)$. In effect, it uses expected utility analysis given a partition of worlds to obtain $U(p)$ and then Section 2.2.2's form of intrinsic utility analysis to compute the utilities of worlds from the intrinsic utilities of objects of BITs realized in them. As Section 5.2.2 shows, the hybrid is viable.

Now consider a BIT by BIT intrinsic utility analysis of $U(p)$. Although it uses chances for realizations of BITs, it is not a hybrid of expected utility analysis and another form of intrinsic utility analysis. It is independent of expected utility analysis because it involves no partition of states. A partition of states is unnecessary because reliance on the set of BITs ensures that relevant considerations are neither omitted nor double-counted.

As a final illustration, I construct a viable hybrid utility analysis of $U(o)$ by starting with an expected utility analysis of $U(o)$ with respect to states independent of the option and then using intrinsic utility analysis to obtain the utility of the option given a state. For brevity, I focus on the second step. That step obtains $U(o$ given $s)$ using a

tain, it uses a mean evaluation of its possible courses. Under common assumptions, it generates a mean-risk ranking of options. For this approach, see Weirich (1987).

Partial utility analysis justifies mean-risk utility analysis more simply and less controversially. It focuses more directly on the BITs that are the source of an option's evaluation. Also, it need not resolve difficulties in identifying the time at which a BIT is realized. It need only consider whether a BIT would be realized if an option were realized.

conditionalized BIT by BIT intrinsic utility analysis. It takes the state s as an assumption, indicatively supposed, and relativizes the analysis to that assumption. That is, it applies this formula:

$$U(o \text{ given } s) = \Sigma_j P_o \text{ (BIT}_j \text{ given } s) IU(\text{BIT}_j),$$

where BIT_j ranges over objects of BITs. The formula relativizes probabilities to s, but it does not relativize intrinsic utilities to s because $IU(\text{BIT}_j)$ is unaffected by s's supposition (see Section 2.1.3).

Of course, two hybrid methods of utility analysis may provide two methods of calculating the same quantity. Then one wants to know whether the two methods are consistent. Section A.5 demonstrates the consistency of two of this chapter's hybrid methods.

5.A.2. Generalization of Partial Utility Analysis

Partial utility analysis may be generalized. This section presents a version of Section 5.3's partial utility analysis that uses a partition of BITs rather than individual BITs.

An analysis of $U(p)$ with respect to partial utilities is clearest when the partial utilities correspond to the members of the set of BITs. However, they may correspond to any partition of the set of BITs. The analysis may, for example, use p's partial utility with respect to one BIT and then its partial utility with respect to all BITs save that one. The sum of the two partial utilities equals $U(p)$.

To illustrate the coarse-grained form of partial utility analysis given uncertainty about realizations of BITs, consider a lottery. An analysis may divide its utility into its partial utility with respect to a subset of BITs and its partial utility with respect to the complement of that set. In effect, the analysis divides the lottery's utility into the partial utilities of two lotteries over realizations of BITs. For example, suppose that exactly four BITs exist whose objects are a, b, c, and d. Then consider the lotteries in Figure 5.2, where the chance events are the same for each lottery and the outcomes are realizations of BITs. The utility of L is the sum of the utility of L_1 with respect to the BITs toward a and c, and the utility of L_2 with respect to the BITs toward b and d.

Allowing partial utilities to be taken with respect to any partition of BITs, the general form of partial utility analysis advances this formula:

$$U(p) = \Sigma_j U_j(p),$$

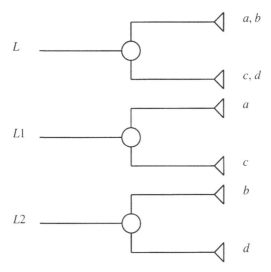

Figure 5.2. Lotteries for BITs

where U_j is utility with respect to the *j*th member of a partition of BITs. The same type of generalization works for the hybrid form of analysis combining partial and expected utility analyses, and for conditional partial utility analysis. Also, a partial utility taken with respect to a set of BITs may itself be analyzed using a partition of the set, and partial utilities with respect to its members. For example, to obtain the utility of a proposition with respect to three BITs drawn from a larger set of BITs, we may add the proposition's utility with respect to the first two BITs and its utility with respect to the third BIT.

6

Group Utility Analysis

Section 1.1's analysis of a new safety standard's utility for society entertains its division according to people, a traditional dimension of utility analysis. Combining an analysis along this dimension with analyses along the dimensions of goals and possible outcomes yields a three-dimensional analysis of the safety standard's social utility. The standard's utility for a person, a point on the dimension of people, may be expanded along the dimension of goals, as in Chapter 2. To accommodate uncertainty, it may be expanded simultaneously along the dimension of possible outcomes, as in Chapter 4. The three dimensions provide a versatile framework for analyses of social utility.

To analyze an action's group utility with respect to people in the group, one has to decide how to add the action's benefits and costs for those people. If a society adopts a new safety standard, some people may benefit because of a reduction in injuries. Others may not benefit personally but still share the new standard's cost. How should the analysis weight these considerations so that they add up to the standard's social utility? This chapter advances a procedure. It argues in a utilitarian vein that an action's group utility is a sum of its interpersonally scaled utilities for the group's members. However, because the procedure tries to assess a group action's instrumental choiceworthiness, that is, the extent to which the group action serves group goals, it weights a member's utility by his power in the group before summing. It advances a type of weighted utilitarianism. According to to it, *an action's group utility is a power-weighted sum of its utilities for the group's members.* This chapter elaborates and supports this account of group utility analysis, and shows how to combine it with intrinsic utility analysis and expected utility analysis. Because I work within the school of thought that accepts the possibility of interpersonal utility and quantitative power, I provide only brief introductions of each.

My account of group utility is designed for a theory of rational group action that recommends maximizing group utility. The principle to maximize group utility, because it aims at instrumental rationality only, has more modest ambitions than its counterpart in utilitarian moral theory. It does not aspire to be a principle of right action. To win support as a principle of right action, it needs, as a minimum, the assumption that group utilities derive from moral personal utility assignments and justified differences in power. Moreover, even under the additional assumption that weights for group members must be equal to be justified, its credentials are still better as a principle of beneficence than as a principle of right action.

Because I take maximizing group utility as a principle of instrumental rationality only, I bypass criticisms of utilitarian moral theory. I also bypass other criticisms resting on principles of morality or noninstrumental rationality. It is undeniable that a group's actions are influenced by differences in its members' power. It is also clear that its members' rational exercise of their power is part of the group's rational actions. A group's actions prima facie ought to meet standards of instrumental rationality and also other standards, including standards of morality and noninstrumental rationality. I do not address those other standards, however, or explain how to resolve conflicts of standards. My normative theory is narrowly focused on instrumental rationality in ideal cases. Its objective is merely to show how in at least certain ideal cases one may treat groups as agents to which familiar standards of instrumental rationality apply.

6.1. GROUP UTILITY ANALYSIS'S ROLE

Before presenting group utility analysis in detail, let me sketch some applications to motivate it. Suppose that a pair of friends, Marina and Pierre, want to have dinner together at a restaurant. Marina prefers an Italian restaurant, whereas Pierre prefers a French restaurant. Their custom is to settle their differences by adopting the preference of whomever contributes more to the dinner fund, the proceeds of which defray equally the costs of their dinners. Marina is wealthier than Pierre and so may contribute a larger sum of money before expending a unit of interpersonal utility. If each gains a unit of interpersonal utility from dining according to preference, Marina may keep one-upping Pierre until he concedes, without reaching the point where the utility of the money she contributes exceeds the utility of winning her preference. If they both realize this, Marina may make a minimal contribution to the dinner fund, and Pierre may decline to compete. Then they go to the Italian restaurant.

In this example the pair's action depends on the power of its members. Because Marina's wealth makes her more powerful than Pierre, she has more control than Pierre of the pair's action. If we put power aside, the possible actions tie. The pair derives the same utility, the same sum of interpersonal utilities, whichever restaurant it selects. Marina's greater power, however, tips the balance toward her preference. Going to the Italian restaurant maximizes the pair's utility. Marina's greater power may tip the balance without being exercised. The pair's recognition of it may suffice.

Adding a third member to the group may not substantially change group dynamics. Suppose that Juan joins Marina and Pierre, and he prefers a Mexican restaurant. Each gains one unit of interpersonal utility from dining at a preferred restaurant, and each is indifferent between the preferred restaurants of the other two. Although a coalition may form, it still has to settle its preference after purchasing the right to be decisive. Suppose, for simplicity, that the bigger contributor to the coalition will settle its choice. In these circumstances coalitions have no incentives to form. Ability to form coalitions does not affect a member's power. If the friends promote their preferences by contributing to the dinner fund, the wealthiest will win. A rational group action favors the most powerful, and that remains the wealthiest.

A rational group action is a matter of power-weighted utilities only in cases where interpersonal utilities exist and the power of members is quantitative. Only ideal cases meet these conditions. Much of interest to game theorists is compressed into the measures of members' power in their group. A member's power is just a summary of his degree of control over group action. In cases where a single power rating for each member misrepresents group dynamics, my method of group utility analysis does not apply. My idealizations and restrictions target cases, such as the example, where interpersonal utilities and power ratings summarize in an abstract way the essential features of rational group action. My method of group utility analysis is designed for such cases. As confirmation of it, Section 6.5 shows that in ideal cases rational bargaining within a power structure maximizes group utility as I define it.

In realistic cases, where interpersonal utilities and quantitative power do not exist or are hard to assess, will group utility analysis be relevant? Applications may rely on approximation to make headway. Group utility analysis may still guide prediction and explanation. Suppose a theorist wants to predict or explain a group's action. Because in ideal cases a group maximizes group utility, in cases approximating the ideal, options' group utilities bear on the group's action. Because group utility analysis advances a principle of rationality for groups, its relevance for prediction and explanation depends, of course, on the extent to which the group's members are rational and conditions are conducive to the group's acting as a rational agent. Nonetheless, because people are by and large rational and because groups, given opportunities for communication and coordination, approximate rational agents, in some cases group utility analysis identifies factors, such as differences in power, behind a group's actions.

For example, in some groups a leader has authority that other members lack. His views have more weight than others' views. This is so in a tribe ruled by a chief. If a historian were explaining the tribe's actions, weighted utilitarianism would provide a convenient framework. The chief, having more weight than others, steers the tribe toward actions he favors. One expects tribal action in accord with his wishes.

Weighted utilitarianism may also guide predictions of actions taken by a country's parliament. Suppose that the parliament is divided according to political party. For simplicity, imagine that a party's power-weight goes by the number of seats it holds and that its interpersonal utility assignment to policies goes by their centrality in its platform. If the parliament decides on a partisan issue, and coalitions are static, weighted utilitarianism may approximately gauge the parliament's action.

Also, consider predicting the actions of an international alliance. A nation's weight in the alliance may be approximated by the size of its population. Its utility assignment may be approximated by public opinion polls. If coalitions among the alliance's members are static, weighted utilitarianism may estimate the alliance's course of action. Weighted utilitarianism offers an alternative to the predictive model of Bueno de Mesquita (1981). He assumes, for simplification, that a nation functions as a dictatorship in its international relations, with the preferences of an elite overruling all others. Weighted utilitarianism handles cases like this, but also cases in which national power is divided. It can obtain a nation's preferences from an assessment of individuals' power and goals. It provides a more versatile framework for predicting an alliance's behavior.

Approximate knowledge of group utilities serves purposes besides prediction and explanation. For example, a group member may acknowledge that some members, because of their seniority, have more power than others. If out of solidarity he wants to follow the group's desires, he may then use weighted utilitarianism to assess them. Similarly, a member of a business partnership may recognize that the partner with the greater investment has a greater voice and may use weighted utilitarianism to calculate the partnership's utility assignment. When acting on behalf of the partnership, she may want to promote its utility defined in terms of the difference in power. A corporation's CEO, in fact, has an obligation to act on behalf of the shareholders. She may weight the shareholders' goals according to the number of shares held and then maximize utility for the corporation as defined with respect to that weighting. The next chapter treats such trustee decisions in depth.

Because group utility analysis shows which group actions are rational, it has an important role in political philosophy. It contributes to the justification of group actions in both ideal cases and more realistic approximations. Group utility has a prominent place in democratic theory, for example. It represents the will of the people, the justification of a state's actions according to Rousseau. If a citizen wants to vote in accordance with the will of the people, she needs a way of assessing it. A thoroughgoing democrat may want to respect the will of the people even when it is mistaken. Some ways of explicating the will of the people are egalitarian. But if one explicates the will of the people in terms of the people's rational actions, differences in power need accommodation. Weighted utilitarianism yields a notion of the will of the people that reflects political power and interpersonal utility.

Group utility also has a theoretical role in the systematization of principles of group rationality. The literature on social choice, for example, proposes the standard of Pareto optimality. According to it, a group's choice is rational only if no alternative is Pareto-superior – that is, better for some members and at least as good for all according to their personal utility assignments. The literature on social choice also proposes a principle of collective rationality that prohibits a cycle of choices among alternatives, say, rejecting a for b, then b for c, and finally c for a. One way of systematizing principles of group rationality is to derive some from others. Satisfying the principle to maximize group utility entails satisfying the principles of Pareto optimality and collective rationality. In cases where options have group utilities, maximization of group utility supports these principles of group rationality. The intuition that rational groups act to maximize group utility thus helps explain the intuitions that they act to achieve Pareto optimality and to avoid intransitivities of choice.

6.2. GROUPS AS AGENTS

Let me now begin laying the groundwork for a detailed presentation of group utility analysis. I consider the nature of group actions, to which group utilities attach, and the grounds for standards of rational group action, such as maximization of group utility.

Some collections of people we call groups and regard as agents that perform actions. These include nations, corporations, clubs, and committees. Because we regard them as agents, we apply to them the standards of rationality for agents. For example, we believe that their actions should be governed by their desires. The application of standards of rationality to groups is not straightforward, however. Groups do not have minds, so attributing desires to groups calls for a technical definition of a group desire. The goal in formulating such a definition is to introduce a useful concept of group desire, one whereby standards of rationality apply to groups and, in particular, group desires govern rational group actions. This chapter defines group desires quantitatively in cases where they are rational and formed in ideal conditions, and assigns group utilities to their objects. Because group desire is an artifact, its treatment in nonideal cases raises issues too complex to address here.[1]

I define group utility in terms of the utility assignments of the group's members. It is not a primitive, undefined theoretical entity. So a group utility analysis is just an application of group utility's definition. The definition's direct application to a group is the fundamental form of group utility analysis. Derivative, equivalent forms of group utility analysis may divide the group into subgroups and break down utility for the group into utilities for the subgroups. Because utilities for the subgroups depend on utilities for their members, the group's utility still depends on its members' utilities.

My definition of group utility applies to idealized group decision problems. To make uncertainty manageable, my idealizations stipulate

---

[1] Earlier treatments of this chapter's topic are in Weirich (1990, 1991a, 1991b, 1993).

that the costs of members' sharing information can be compensated and that shared information generates a unique probability assignment. Given such idealizations, I define group utility as a power-weighted average of members' interpersonal utilities. This definition allows for members' having conflicting goals and makes group utility a compromise utility assignment. I show that maximizing group utility so conceived yields a rational group decision, one achieved by rational individual decisions in ideal conditions for group action, even where uncertainty calls for application of expected utility analysis along with group utility analysis.

A group decision must be technically defined because groups do not have minds and so do not literally form intentions. I take it as the realization of a group action, one constituted by the members' actions. The options in a group decision problem are therefore group actions. Intrinsic and expected utility analysis have been formulated to apply to all propositions and so to options taken as group actions. They are not restricted to options taken as an individual's possible decisions. To accommodate the reasons that led Section 3.2.1 to take options as possible decisions and to avoid aggregation of individuals' desires across time, I take a group's options in a decision problem at a time to be group actions the members may realize by simultaneous action at the time – for example, reaching an agreement.

My goal is a definition of group utility whereby a rational group action maximizes group utility, given ideal circumstances for group action. More generally, I want to define group utility so that all of utility theory applies to group action – not just rules for maximizing utility, but also rules for maximizing expected utility when conditions are ideal except for uncertainty. Furthermore, the definition of group utility should minimize limits on the application of utility theory to groups. I want utility theory to apply to as wide a range of cases as possible.

To meet this goal, one cannot define group utility arbitrarily and then simply adjust standards of rational group action so that the actions meeting the standards maximize group utility. At least one standard of rational group action is independent of the definition of group utility and the application of utility principles for agents to groups. It states that a rational group action issues from rational actions by all the group's members. Hence my definition of group utility must yield the result that in ideal cases a group maximizes group utility if all members act rationally.

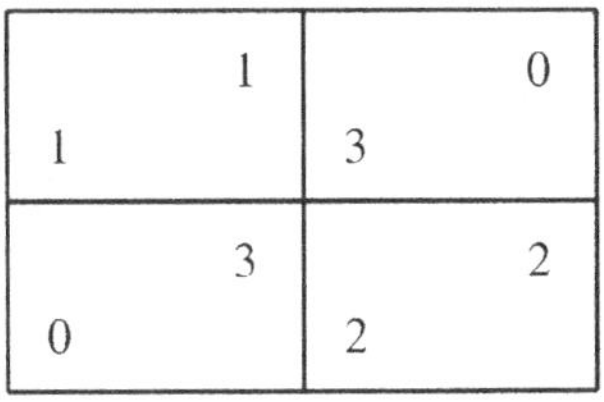

	1	0
1	3	
	3	2
0	2	

Figure 6.1. The Prisoner's Dilemma

I assume that group rationality reduces to the rationality of the group's members. The rationality of a group's action is ultimately to be explained in terms of the rationality of its members' actions. Even principles of group rationality such as the principle of Pareto optimality must ultimately be explained in terms of the rationality of the members' actions. The rationality of a group's action cannot be explained ultimately in terms of group desires because those desires are not primitive theoretical entities but are defined so that in ideal conditions rational group actions are governed by group desires. Group desires achieve only intermediate-level explanations of rational group actions.

The compositional standard of rationality for group action may seem wrong in cases such as the Prisoner's Dilemma, depicted in Figure 6.1. Rows represent options for one agent and columns represent options for another agent. In a box, which represents a combination of options, one for each agent, the number in the lower left-hand corner is the utility gain for the agent choosing the row, and the number in the upper right-hand corner is the utility gain for the agent choosing the column. The two agents cannot communicate about the actions to perform and cannot bind themselves to a joint strategy.[2]

In the Prisoner's Dilemma it may seem that rational actions by individuals produce an irrational action by the group. Because the top row and the left column are dominant, the group action produced by that combination is rational according to the standard of composition by rational individual actions. Yet the bottom row and right column produce a group action that is better for each member, and so Pareto-superior. It appears to be the rational group action. Appearances are deceptive, however. The top left box represents the rational group

[2] There is an extensive literature on the Prisoner's Dilemma. For an introduction to it, see Campbell and Sowden (1985: pt. 2).

action even though that action falls short of the standard of Pareto optimality and the standard of maximization of group utility (on any plausible definition of group utility), because circumstances, in particular, the inability of the individuals to communicate and make binding contracts, provide excuses for the group's failing to meet those standards. Those standards are not mandatory in nonideal circumstances. In general, circumstances may provide excuses for a group's failing to meet the standards of rationality that apply to it taken as an agent. A group's action may therefore be rational in virtue of meeting the compositional standard, although it falls short of other, more lofty standards of rational group action. Those other standards may express ideals of conduct rather than necessary conditions of rationality for group actions undertaken in nonideal circumstances.[3]

Ordinary usage does not call a collection of individuals a group unless it can attain some goals of rationality, and so qualifies as an agent in virtue of having some obligation to attain those goals. For simplicity, I depart from the usual view of groups. I take any collection of individuals as a group. I call it a group even if it is not an agent and is not in a position to meet even basic standards of rationality. On my view, goals of rationality always apply to collections of individuals but sometimes obstacles stand in the way of their attainment. Sometimes a collection of individuals fails to reach goals of rationality it is perfectly able to attain and then is an agent, but an irrational one. Other times it fails to attain basic goals of rationality because it cannot attain them. Then it is not an agent, and so is not a group on the ordinary view, but nonetheless is a group on my view, and is rational because attaining the goals is beyond its abilities.

One of the chief agent-making (and in the ordinary sense, group-making) circumstances of a collection is the ability of its members to communicate with each other. We do not, for instance, typically regard collections of scattered individuals as groups that act, because communication is impossible. Pioneer migration westward as a result of settlers' scattered, uncoordinated acts to gain elbow room is not usually regarded as a group action. Another circumstance that contributes to a collection's status as an agent is an ability to organize, for instance, to make binding contracts (even if the only binding force is honor). A group as ordinarily understood is a collection of individuals able to

---

[3] Binmore (1994: 103) similarly attributes Pareto inefficiency in the Prisoner's Dilemma to the agents' nonideal circumstances.

achieve some coordination – for example, agreement about a course of joint action. However, I regard any collection of individuals as a group, whatever its circumstances, even given insurmountable barriers to communication and coordination.[4]

In special circumstances a group's members may be able to coordinate their acts without agreements. Perhaps they are playing a game with several equilibria, one of which is optimal. Then given their rationality and an ability to anticipate each other's acts, they may achieve the optimal equilibrium without an agreement. Or perhaps custom establishes a convention for a group. Without an agreement, each member conforms expecting others to conform. Drivers in the United States, for instance, stay on the right of the road following a convention rather than an agreement. Although agreement is not necessary for coordination, my idealizations provide the means of achieving coordination through agreement. Communication and binding agreements generate anticipation and coordination of acts in a simple, direct way that is appealing from a theoretical standpoint.

When circumstances favor in every way a collection of individuals' acting as an agent – moreover, as a rational agent, so that the collection is perfectly able to attain the goals of rationality for agents – I say that circumstances are ideal for group action. These ideal circumstances include not only circumstances that permit group action in the ordinary sense but also circumstances that facilitate it, in other words, make it easier for the collection of individuals to act as an agent and attain the goals of rationality for agents, as the idealizations for individuals do (see Sections 1.3.2 and 3.2.5). They include the members' being in ideal conditions for individual rational action, the opportunity for costless communication between the members, the opportunity for cost-free binding contracts, and, in general, the opportunity for unconstrained negotiation.

Some ideal conditions – for example, those facilitating communication – concern the group's ability to act as an agent in the ordinary sense. Because of the connection between a group's acting as an agent and the applicability of standards of rationality to it, these ideal conditions facilitate meeting standards of rationality as well. In contrast, some ideal conditions concern only the group's ability to meet standards and goals of rationality, for example, certainty about the

---

[4] For an account of the features of groups that make them agents, see Gilbert (1996: chap. 6).

177

outcomes of its actions. The first set of ideal conditions concerns the group itself, taken as an agent, and the second set concerns the group's situation.

I hold that the rational actions of a group conform to the standards of rationality for agents when conditions are ideal for group action. When circumstances are not ideal for group action, a rational group action may fall short of the standards. The group action may be the result of rational actions by the group's members and yet fall short of the standards for rational group action because of some impediment to the group's acting as an agent. Or the group's circumstances may create obstacles to meeting goals of rationality. Then there may be some secondary goals that the group must reach to be rational, given the obstacles to reaching the primary goals. For instance, a group may be perfectly able to act as an agent, but the members may be uncertain about the outcomes of group actions. Then given the obstacle to realizing the possible outcome of maximum group utility, the group should maximize expected group utility.

Conflicting preferences among the members of a group do not hinder the group's acting as an agent, at least not given the rationality of its members. Defining group utility to treat the group as an agent simply requires a method of aggregating diverse private goals to obtain group goals. We may conceive of conflicts between members along the line of conflicts between an individual's desires. The aggregation of individual utility assignments to form a group utility assignment is like the aggregation of an individual's conflicting desires to form his desires on balance.

The idealization that removes individuals' uncertainty about outcomes of actions upgrades a group's chances for success rather than its ability to act as an agent. Conditions may be ideal for a group's acting as an agent despite uncertainty. Uncertainty bears on the standards of rationality appropriate for the group rather than the group's ability to meet those standards. Uncertainty about the outcomes of actions makes it appropriate to evaluate the rationality of a group action in terms of the goal of maximizing expected group utility, a secondary goal for pursuit of the primary goal of obtaining the outcome at the top of the group's preference ranking. Although uncertainty may prevent the group's obtaining the outcome it prefers, uncertainty does not prevent the group's meeting the standards of rationality applicable in its circumstances when taken as an agent, and so does not hinder its acting as an agent.

178

Besides idealizations that remove obstacles to goals of rationality, I impose some idealizations that remove factors complicating the justification of real-life group decisions. To remove the complication of imprecision, for example, I assume a foundation for quantitative group utility. I want to apply utility theory to groups, with the understanding that the parts of utility theory calling for quantitative utilities are limited to cases where these quantities exist.

Having considered the objectives of a definition of group utility, I now review the methods of definition available. Because group utility is not a primitive used to explain rational group action but rather a construct used to represent rational group action, one is free to define group utility in terms of group preferences without loss of explanatory power, in contrast with the case for an individual's utility assignment (see Section 3.A.1). Nonetheless, there is a serious problem with this approach. To carry it out, one first defines group preferences in terms of hypothetical group choices, that is, hypothetical group actions. Then one tries to obtain group utilities from these group preferences. For this, one must know group preferences concerning many pairs of actions, and so, choices, perhaps hypothetical choices, from a variety of sets of options. In the literature's terminology, one must know the group's *choice function* from sets of available options (taken from a *universal set* over which utility is to be defined) to choices to realize a particular option or, in case of ties, choices to realize one option in a particular set of options. But a group's choices are sensitive to the set of options available, for the options available determine the negotiating advantages of the group's members. Therefore one cannot expect choices from various sets of options to reveal the group's preferences among the options in the universal set.

To illustrate, suppose that two agents undertake to divide $100; they must agree on a division to obtain any of the money. Also, suppose that for each the utility of an amount of money gained is proportional to the amount. The triangle in Figure 6.2 represents in utility space the possible resolutions of their bargaining problem. Suppose that the pair settles on an equal division of the money. The midpoint of the hypotenuse represents this outcome. Next, imagine that agent 2 faces a public, external constraint. She may have no more than half the money. The trapezoid in Figure 6.3 represents in utility space the options available in this derivative bargaining problem. The set of options available includes all the options of the first problem except those represented by the small triangle omitted. Suppose that the pair

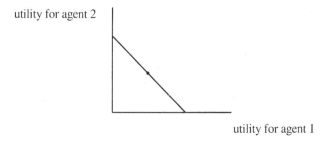

utility for agent 2

utility for agent 1

Figure 6.2. A Bargaining Problem

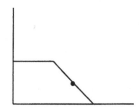

Figure 6.3. A Truncated Bargaining Problem

settles on a 75–25 split. The dot in Figure 6.3, not the midpoint of the original triangle, represents this outcome. A rational pair of agents in ideal conditions might reach this nonegalitarian outcome because the change in options available entails a loss of negotiating power for agent 2. However, no rational preference ranking of options in the universal set (we can suppose that it is the original triangle) agrees with the preferences revealed by these two group actions, under the assumption that preferences do not change with the options available. If the midpoint of the original triangle's hypotenuse is preferred to all other options in the universal set, then because it is still a member of the truncated set of options, it must be preferred to all other options in the truncated set if preferences are to be rational.

In light of such examples, I reject the idea that group preferences are revealed by choices from various sets of options. Group preferences, defined in terms of group choices, are too sensitive to changes in the options available. The underlying preference ranking is not revealed by a series of choices made in different contexts.[5]

---

[5] In more technical terminology, I reject Nash's (1950) independence condition and I reject Sen's (1970: 17) alpha and beta conditions. See Luce and Raiffa (1957: 128–34) for some criticism of Nash's independence condition. This criticism also

180

One might try to resolve the problem by revising the method of obtaining group utility. Instead of using group choices to reveal preferences, one might start with preferences. Instead of asking for a choice from some set of available options, one might ask for the top preference in some set of options taken abstractly and not as the set of options available. Such preferences are all with respect to the same circumstances. The preference ranking of options does not change because the group's circumstances, real or hypothetical, do not change as deliberations move from one abstract set of options to another. The price of this procedure is a definition of group preference independent of group choices; one cannot take group preference as primitive because groups do not actually have minds. However, the only way to define group preference independently of group choice so that it fulfills its theoretical role (governance of rational group action) is to define group utility first and then group preference in terms of it. Defining group preference in terms of the preferences of the group's members is prohibited by Arrow's (1951: chap. 5) possibility theorem (see Section 6.A.1). Consequently, there does not seem to be any way to extract group utility from group preferences. The next section uses a different method to define group utility. It relies on interpersonal utility, one of the conceptual resources contextualism affords.

6.3. GROUP UTILITY

I define utility for a group in terms of features of the group's members, principally their utility functions after these have been put on the same scale, and so transformed into interpersonal utility functions. Controversy surrounds interpersonal utility. For my purposes, interpersonal utility's conceptual possibility suffices. I apply group utility analysis only to ideal cases where interpersonal utilities exist. I do not claim that utility is interpersonally comparable in every case. Perhaps the utility one person attaches to a Cezanne landscape is not comparable with the utility another person attaches to famine relief. Then I put aside this case. Utility does not have to be interpersonally comparable in all cases to be interpersonally comparable in some cases, and

applies to Sen's alpha and beta conditions; alpha is Nash's independence condition. See also Eells and Harper (1991), Vallentyne (1991), Vickers (1995), Rawling (1997), and Weirich (1998: chap. 4).

one may restrict group utility analysis to cases where it is interpersonally comparable.

Some theorists doubt the meaningfulness of interpersonal utilities. They deny their very possibility. I do not review this issue. As in utilitarian moral theories, I assume that interpersonal utilities are meaningful even if not operationally defined. In fact, people routinely make interpersonal comparisons of utility using behavioral and biological evidence. They may, for instance, conclude that Jones wants a certain job more than Smith does because Jones tries harder than Smith to win the job. Such everyday comparisons ground the concept of quantitative interpersonal comparisons of utility.[6]

Let me defuse one common objection. Interpersonal utility is often rejected on the grounds that minds are private and so incomparable. At first this appears to be an argument that interpersonal comparisons of utility are beyond the limits of human knowledge, not that comparative relations fail to apply interpersonally to utilities. But the argument takes the privacy of minds in a metaphysical sense so that the inapplicability of the comparative relations follows. The argument from the privacy of minds, however, rests on a dualist conception of mind. Given a materialistic (or physicalistic) conception, minds are no more private than other bits of matter. In particular, the properties of minds can be duplicated in principle (even the qualia experienced). It is possible for two people to be in the same general mental state. Suppose then that two people are psychological duplicates in duplicate environments. Imagine that the first person wants to gain $2 and wants to gain $1, and that the first desire is twice as intense as the second. Then the second person has the same desires with the same intensities. Consequently, the first person wants to gain $2 twice as much as the second person wants to gain $1. Granting a materialistic conception of the mind, minds are comparable.

This materialist argument for utilities' interpersonal comparability does not suppose that utilities register only pleasures or feelings of sat-

---

[6] Even if an operational concept extensionally equivalent to interpersonal utility is found, it will not fill the explanatory role of interpersonal utility. It will not get behind the phenomena, the actions of a group, to explain them. For a defense of interpersonal utility, see Weirich (1984a) and Goldman (1995). For a survey of economists' ideas about interpersonal utility, see Binmore (1998: 168–78, 381). For an argument against defining interpersonal utility in terms of extended preferences about living the lives of others, see Broome (1999: chap. 3). His argument treats utility as a measure of the good but carries over to utility in my sense.

isfaction. Utilities may depend on agents' beliefs and desires in complex ways and yet be interpersonally comparable because they are ultimately reducible to agents' physical states, which are comparable.

Also, let me emphasize that for my utility theory's purposes, interpersonal utilities need not exist in the real world. They are needed only in the ideal cases to which the theory applies. All that matters is whether interpersonal utilities are conceptually possible. Given that they are, one may nonvacuously limit group utility analysis to the ideal cases in which they exist.[7]

In defining group utility, I also use the power of individuals in the group. The kind of power of interest is sometimes called social power to emphasize that it is power over people rather than over animals, inanimate objects, and events such as the weather. More precisely, I am interested in just one variety of social power, economic power. This type of power is called economic because it affects behavior through the provision of incentives. It excludes brute force. It is not exercised, for example, by carrying sleeping people, unmoved by incentives, to new locations. However, I understand economic power in a broad sense so that incentives need not be monetary or involve commodities. Economic power may arise, for example, from authority as well as wealth and may involve incentives that concern approval.[8]

I distinguish power from influence. Power works by incentives, whereas influence works by persuasion. Power over another's choice is an ability to change the options, states, or consequences that are relevant to his choice, whereas influence over another's choice is an ability to change his utility assignment for those options and consequences, or his probability assignment for those states. Under the idealizations, individuals are immune from others' influence on their opinions because they are rational and know that others have the same information they do.

I also distinguish power from advantage. An advantage is anything that improves one's prospects. Power is a means of improving one's prospects by affecting others' incentives. An advantage is enjoyed but

---

[7] My argument for utility's interpersonal comparability uses premisses stronger than necessary. It must establish only the possibility of two agents with some duplicate quantitative desires. Premisses weaker than materialism suffice for this. Some metaphysicians who reject materialism, such as Jackson (1986) and Chalmers (1996), may nonetheless concede the possibility of agents with duplicate desires.

[8] See Barry (1976) on economic power. For an account of social power in general, see Goldman (1972).

not necessarily exercisable, whereas power can be exercised. Someone with an advantage may be better able to exert pressure. But only the ability to exert pressure is power. For example, resources are an advantage in every case but provide power only in situations where they can be used to affect others' incentives. A person's indifference to risk, or ability to wait for an agreement, may be an advantage in bargaining with other people. But these factors are also sources of power only if they can be exploited to affect others' incentives.

This account of power is a brief introduction. A fuller account would more precisely classify types of power; would further articulate distinctions between power, influence, and advantage; and would explain the origin of power from factors such as wealth. For my purposes, however, a brief introduction suffices. In examples, where I need more precision, I add supplementary details. In particular, I impose the restriction that a person's ability to provide incentives depends exclusively on the utility cost for him of providing utility gains for others. In the cases I treat, an individual's power is given by the rate at which he can transfer utility from himself to others. The less it costs him to transfer utility to others, the more power he has over them. Section 6.5 elaborates these assumptions about power.

Group utility need not be defined in all cases because it is just an artifact introduced to enable application of utility theory to group actions. My definition of group utility is not advanced for all groups in all circumstances. It is advanced only where interpersonal utility and quantitative power exist. The definition's assumptions form explanatory idealizations reducing the complexity of explanations of group actions. Although interpersonal utility and quantitative power may exist in ideal cases only, the results of group utility analysis in ideal cases are relevant to realistic cases. In particular, they justify rules of thumb for rational group action in the way that utilitarianism, granting its accuracy in ideal cases, justifies majority rule as a practical means of approximating the moral course of action.

The group utility of a proposition – or, in decision problems, an option – is defined following the principle of pros and cons. I begin with a separation of considerations for and against the proposition and then attach a group utility to each. The group utilities of the considerations are added to obtain the group utility of the proposition. The fundamental step is the definition of the group utility of a consideration, the proposition's outcome for a member.

To separate considerations, I use the utility of the proposition for each member of the group. These utilities for individuals are pros and cons according as the utilities represent gains or losses. Next, each individual's utility is weighted by the individual's power. The group utility of a utility gain or loss for a member depends on the member's power in the group. The power-weighted utility yields the group utility of the consideration. Finally, I add the power-weighted utilities to obtain the proposition's group utility. Thus, using o to stand for a group option, GU to stand for group utility, and e_i to stand for the economic power of the ith member in an arbitrary order of the group, the following definition emerges:

$$GU(o) = \Sigma_i e_i U_i(o).$$

To motivate this definition, I appeal to the principle of pros and cons, taking group utility informally as a measure of an option's attractiveness to a group. My definition's separation of considerations does not omit anything relevant given that group utility is a matter of individual utilities only. Every group member's utility assignment has a role in settling GU's values. Also, the definition's separation of considerations does not double-count any factor affecting group utility. GU counts a group option's outcome for an individual only once.

To support these claims further, consider omissions. Some may think that the formula for group utility omits global features of the utility distribution produced by an option, such as equality. It does not; it counts these consequences when they are relevant. In applications of utility theory to rational group actions, global consequences should affect group utility only if some members care about them. But in that case they affect the utilities of the option for members; the option's utility for an individual considers everything the individual cares about, including global properties of the option's utility distribution.[9]

Given that an option's utility for one individual may affect the option's utility for another individual, the existence of an option's group utility requires well-behaved dependency relations. Given interaction between utilities for individuals, a group utility assignment presumes an equilibrium among members' desires. In some cases equilibrium may not be reached so that no group utility assignment exists.

---

[9] See Broome (1991: chap. 9) on individualistic egalitarianism.

For example, imagine a two-agent group deciding between two options. Agent 1 wants whatever option agent 2 opposes and agent 2 wants whatever option agent 1 wants. The agents are incapable of indifference between the options. In this case the options' group utilities are undefined because their utilities for the group's members are undefined. This result does not undermine the definition of group utility. It shows only that the definition does not apply to all cases.[10]

Now consider double-counting. Some may think that if one member of the group cares about the utility level of another (because of envy, prejudice, or altruism, say), then the utility level of the second member is counted twice, once by his own utility function and once by the first member's utility function. The apparent double-counting is not genuine, however. The same consideration is not counted twice. Rather there are two related considerations, each counted separately. The first consideration is the utility level of the second member, and the second consideration is the utility level of the first member as influenced by the utility level of the second member. Utility for a member takes account of all sources of utility.

To clarify the definition of group utility, let me compare it with the utilitarian definition of social utility. Utilitarianism defines social utility as a sum of individual utilities. If o is a social option, its social utility $SU(o)$ is $\Sigma_i U_i(o)$, where $U_i(o)$ is o's utility for the ith member in an arbitrary order of the society. Given this definition, utilitarianism says to maximize social utility. The utilitarian definition of social utility counts every person as one and none as more than one. That is, it gives the utilities of society's members equal weight. Weighting individuals equally is appropriate within the framework of a moral theory but inappropriate within the framework of a theory of rational group action. It is rational for the members of a group to use their power to steer the group toward actions they want. The influence of a rational individual on a group action is a product of the intensity of her

---

[10] As the example shows, where the utility of an action for one person depends on its utility for others, utilities may be ungrounded. Cases of ungrounded utilities are similar to cases of ungrounded truth values for paradoxical sentences such as these:

(A)   Sentence (B) is true.
(B)   Sentence (A) is false.

In cases where utilities exist despite feedback loops, they are grounded in an equilibrium among the desires of the group's members, just as in cases where sentences referring to one another have truth values, they are grounded in a stable pattern of truth value assignments.

preferences and the power she has to promote them. Given that a rational group action is the result of rational individual actions, a rational group action bears the mark of the most powerful members. Hence, to obtain a definition of group utility such that a rational group action is one that maximizes group utility, members must be weighted according to their power. This is why, in contrast with utilitarian moral theory, I take group utility to be a power-weighted sum of individual utilities. Might does not makes right, but it makes rational group action.

Notice that a rational group action need not actually favor the powerful. Whether it does depends on what the powerful want and what is rational for them to want. If they want evenhandedness, then a rational group action tends to be evenhanded. If they want self-aggrandizement but this is irrational for them to want, then a rational group action does not tend to give them what they want. The idealization I adopt assumes that a group's members have rational desires. Given this, a rational group action favors the powerful only to the extent that they are self-interested and their self-interest is rational.

My definition of group utility applies to all propositions, not just options. This is helpful. The rule to maximize group utility calls for group utilities of options only, but the rule to maximize expected group utility, applied in the next section, calls for group utilities of possible outcomes, too. Also, to prepare for the next section, I define group utility with respect to conditions using conditional power and personal utility:

$$GU(o \text{ given } s) = \Sigma_i e_{si} U_i(o \text{ given } s),$$

where e_{si} stands for the economic power given s of the ith member in an arbitrary order of the group. Conditional group utility so defined appears in calculations of the expected group utility of an option with respect to a partition of states.

A group member's power is relative to a time and group actions may alter it. The utility she assigns to a group action fully registers her beliefs and desires concerning its consequences for her power. Hence the definitions of an option's group utility and conditional group utility need not conditionalize power-weights with respect to options or states. A power-weight indicates the current influence of a group member's utility assignment on rational group assessments of options and options given states. The variable effects of suppositions of options and states show up in the utility assignments weighted.

This section brings more of utility theory to bear on group action. It applies expected utility analysis to group utilities so that the goal of maximizing expected group utility may guide group action in cases where its outcome is uncertain. Then it applies intrinsic utility analysis to group utilities.

6.4.1. Expected and Group Utility Analyses

By design, expected utility analysis may be applied to any type of utility. It requires only that input and output utilities be of the same type. So expected utility analysis may be applied to group utility if one uses group utility for both options and outcomes. First obtain the group utilities of possible outcomes of a group option, given that a state holds if the option were realized, and then use these utilities to calculate the option's group utility, following Section 4.2's procedure.

$$GU(o) = \Sigma_i P(s_i \text{ given } o) GU(o \text{ given } (s_i \text{ if } o)).$$

Expected utility analysis extends to conditional group utilities. Just obtain the group utilities of possible outcomes under the additional supposition that some condition holds, and then use these utilities to calculate the group utility of the option given the condition. As Section 4.2 points out, the appropriate form of supposition for the condition, subjunctive or indicative, depends on whether the condition is an option or a state.

Of course, to apply expected utility analysis to group utility I need a group probability assignment. The preceding formula uses P to stand for this group probability assignment. I must define the group probability assignment so that when a group maximizes expected group utility, it makes rational decisions. There are two ways to proceed. First, I might construct a group probability assignment that compromises among the member's probability assignments, as the group utility assignment compromises among the member's utility assignments. Second, I might pool the information of the group's members and apply inductive logic to obtain a group probability assignment. Because I want to define group probability so that in ideal conditions it yields rational group actions – the results of rational actions by the members – I want group probability to stem from the information sharing that is part of optimal use of group resources. As the group probability assignment I therefore adopt the probability assignment that, given the idealizations, members reach after pooling information and applying inductive logic.[11]

I assume that inductive logic applied to pooled information produces a common probability assignment. Some accounts of subjective probability, rational degree of belief, hold that two rational ideal agents with the same information may reach different probability assignments; there is room for taste in probability assignments. This point seems right for some bodies of information. Nonetheless, in some cases where information is rich, inductive logic supports a unique probability assignment. I do not defend this view about probability and information here. I work within the school of probability theorists that accept it.[12] Accordingly, the assumption that inductive logic applied to pooled information yields a common probability assignment amounts

[11] Lehrer and Wagner (1981: chaps. 2, 4) propose a method of creating a compromise probability assignment from the probability assignments of a group of experts. The method requires each member of the group to adjust his probability assignment in the light of the assignments of others, and then readjust in light of their adjusted assignments, and so on. Lehrer and Wagner show that given certain natural assumptions about the process of mutual adjustment, the individuals eventually reach the same probability assignment. Hence that probability assignment, they suggest, may serve as the group's probability assignment.

Lehrer and Wagner's proposal is attractive in cases where a group must settle on a probability assignment without exploring the evidence on which each member bases his probability assignment. But where communication is unconstrained, it is more reasonable for the group's members to pool their information and then reach a common probability assignment by applying inductive logic to their pool of information. The probability assignment that results makes full use of the information available to the group.

Aumann (1976) proves that if two rational ideal agents make the same probability assignments and later acquire additional information that prompts revision of their assignments for an event, then, if their new probability assignments for that event are common knowledge, those assignments are equal. This demonstration of identical probability assignments does not assume that the additional information is the same for each agent. It relies on each agent's regard for the other's probability assignment. In this respect it is similar to Lehrer and Wagner's proposal. However, the theorem shows that common knowledge of probability assignments makes information about the basis for the assignments superfluous. It shows that agents may pool the import of their information without pooling the details of their information. So under the theorem's assumptions I consider the generation of common knowledge of probability assignments a way of pooling information rather than a way of compromising among probability assignments.

[12] Harsanyi (1967–8) espouses the view that rational ideal agents with the same information have the same probability assignments. As a result, the view is sometimes called the Harsanyi Doctrine. Kyburg holds a similar view, too. He (1974: 247)

189

to the assumption that pooled information is rich enough to legislate a unique probability assignment. Information is not always that rich, but I assume that it is in the cases I treat to facilitate the application of expected utility analysis to group utility.

Letting the group probability assignment be the members' common probability assignment after sharing information also ensures the consistency of principles for calculating group utilities. Although a common utility assignment is not necessary for consistency, a common probability assignment is necessary. An example brings this out. Imagine a two-person society consisting of Hawk and Dove. This society is on the verge of war with another society. Hawk wants war, Dove wants peace. Each has the same power and intensity of desire. Therefore, with 0 as the utility of the status quo for each, and with 1 as the utility of war for Hawk, and -1 as the utility of war for Dove, the group utility of war is $e_H U_H(\text{war}) + e_D U_D(\text{war}) = (1 \times 1) + (1 \times -1) = 1 + (-1) = 0$. Similarly, the group utility of peace is $e_H U_H(\text{peace}) + e_D U_D(\text{peace}) = (1 \times -1) + (1 \times 1) = -1 + 1 = 0$. Now suppose that a summit meeting is proposed to resolve the crisis. Hawk thinks that the meeting will probably tighten tension and provoke war. Dove thinks that it will probably relieve tension and promote peace. Each therefore favors the meeting over the status quo, and so the meeting has a positive group utility. Because the meeting is in effect a lottery over outcomes with group utilities of zero, this group utility assignment violates the expected utility principle.

Taking a member's chance for an outcome, weighted by his power, as a basic consideration for the group, the first computation is right, the meeting has a positive group utility, whereas the second computation, assigning the meeting a group utility of zero, is wrong. The equation $GU(\text{meeting}) = GP(\text{war})GU(\text{war}) + GP(\text{peace})GU(\text{peace})$ is wrong because there is no group probability function GP in this case. Whatever probabilities are used, the equation does not partition basic considerations. Take the factor $GP(\text{war})GU(\text{war})$. It

thinks that available statistical data, however slender, yields a unique interval-valued probability for a claim assigning a random member of one class to another class. Beebee and Papineau (1997: 223–7) argue that in a wide range of cases information determines unique point-valued probabilities, which they call "relative probabilities." Also, those who endorse a principle of direct inference, such as the Principal Principle of Lewis (1986b: 87), hold that information containing the values of objective probabilities determines a unique point-valued (subjective) probability assignment. Although I may restrict myself to cases where principles of direct inference apply, for generality I treat any case where pooled information is sufficient for a unique point-valued probability assignment.

has to represent the basic considerations $e_H P_H(\text{war})U_H(\text{war})$ and $e_D P_D(\text{war})U_D(\text{war})$. Whatever the value of $GP(\text{war})$, it misrepresents the probability for some basic consideration, and so misrepresents that basic consideration, because $P_H(\text{war}) \neq P_D(\text{war})$. The factor $GP(\text{war})GU(\text{war})$ does not indicate the combined force of the two basic considerations.

The problem with the second computation disappears if Hawk and Dove have the same probability function regarding the possible outcomes of the summit meeting. For any probability p, if for both Hawk and Dove p is the probability of the meeting's provoking war, $GU(\text{meeting}) = e_H U_H(\text{meeting}) + e_D U_D(\text{meeting}) = [p \times 1 + (1 - p) \times (-1)] + [p \times (-1) + (1 - p) \times 1] = 0$. Given a common probability assignment, the group utility assignment satisfies the expected utility principle. $GU(\text{meeting}) = 0$ is the right equation where the agents have a common probability assignment, but not in the original case where they have divergent probability assignments. If group utility analysis is to add group considerations without omission or double-counting, the members must have a common probability assignment. The common probability assignment establishes the separability of group considerations used in the definition of group utility.

A famous theorem of Harsanyi (1955) is pertinent. It says that if individual preferences and group preferences alike satisfy the axioms of a standard representation theorem, and group preferences are Paretian, then the utility assignment representing group preferences is a weighted sum of the utility assignments representing individual preferences. The theorem has implications about the aggregation of individual utility assignments into a group utility assignment given uncertainty on the part of the individuals involved. It entails that if both individual and group utility assignments obey the expected utility principle, and an option of maximum group utility is Pareto-optimal, then an option's group utility must be a weighted sum of its utilities for individuals. In Harsanyi's theorem, however, utilities for individuals are not assumed to be interpersonal utilities. Harsanyi wants to use the derived weights on individual utilities to obtain interpersonal utilities and then utilitarianism. Because the weights in Harsanyi's theorem are uninterpreted, the theorem does not directly bear on my definition of group utility, in which weights represent power. Still, it has an indirect bearing.[13]

---

[13] For discussion of a form of weighted utilitarianism that uses individual utilities weighted according to their worth, see Binmore (1994: 48, 282, 294).

Harsanyi's theorem has the corollary that group utility must stem from a common probability assignment among individuals to be in conformity with the rule to maximize expected utility. More precisely, it follows from Harsanyi's theorem that individual utility assignments that obey the expected utility principle yield a group utility assignment that obeys the expected utility principle and assigns maximum group utility to some Pareto-optimal option only if individuals have the same probability assignment.[14] Now, according to my definition of group utility, options that maximize group utility are Pareto-optimal, if we assume all individuals have at least a little power. For an option that increases utility for some individual and decreases utility for none raises group utility, a positively weighted sum of individual utilities. Therefore, for normal groups, expected utility analysis is applicable to group utility only when all group members have the same probability assignment.

Summarizing, to apply expected utility analysis to group utility, one needs a group probability assignment, and to ensure the Pareto optimality of options of maximum group utility, it has to be a common probability assignment (at least over the propositions that appear in group deliberations). Furthermore, to obtain rational decisions from the application of expected utility analysis to group utility under the idealizations, the common probability assignment has to be obtained by pooling information and applying inductive logic.

My method of obtaining a group probability assignment makes a presumption about information sharing, which I address now. In cases where a group's members are uncertain of its options' outcomes and their information is not uniform, or is "asymmetric," they may be reluctant to share information. Suppose that a group in ideal conditions must make a decision about a matter of common concern in a case where the members are uncertain about the outcomes of the options open to the group. Because, as explained earlier, the idealizations entail that conditions are ideal for group action, and so ideal with respect to agents and situations except perhaps for uncertainty, I assume that the group's members reach a common probability assignment, the one

---

[14] Broome (1987) uses Harsanyi's theorem to show that group preferences conforming to the Pareto principle are guaranteed to maximize expected group utility only if all group members have the same probability assignment. See also Hylland and Zeckhauser (1979), Bordley (1986), and Seidenfeld, Kadane, and Schervish (1989).

obtained by pooling information and applying inductive logic. I assume that in ideal conditions group members, because they are rational, work toward a common probability assignment by sharing information, compensating those who lose an advantage thereby. Because I assume as part of their rationality that members who forfeit an advantage by sharing information are compensated, I must assume that such compensation is possible in the cases I treat. The latter assumption is part of the idealization that conditions are perfect for group action.

To illustrate the idealization, suppose some envisaged sale is called off once the parties pool information. In the short run the potential buyer gains and the potential seller loses from information sharing. Still, in the long run both parties may profit from information sharing. The potential seller, the one who would have profited from the sale in virtue of his additional information, may be compensated by the potential buyer, the one who calls off the sale after information sharing. The potential buyer gains from the information he acquires and may pay for the gain by compensating the potential seller. In ideal conditions for group action, the gains from information sharing are sufficient to compensate those who lose from it. Given the compensation, pooling information is rational for the individuals.

Let me elaborate these points about information sharing. Each member wants his utility assignment to be based on as much information as possible.[15] Given that communication is cost-free, the only barrier to information sharing is the possibility that some group members hold a negotiating advantage in virtue of possessing information others lack. They are not inclined to give away this advantage by sharing information. Under the idealization just reviewed, however, the gains the group derives from information sharing make it possible to compensate them for the loss of their advantage. I assume that rational ideal agents in ideal circumstances for group action work out a means of providing the compensation needed to prompt

---

[15] Good (1967) shows that new information increases the expected value of the option of maximum expected utility and thereby shows that the expected utility of new information is always positive. His theorem explains why individuals want more information, but does not provide a motivation for information sharing. If an individual gives information to his competitors while he receives their information, others' gains in information may decrease his expected expected utility. Information sharing may not boost the expected expected utility of one's choice. For as one agent gains information by sharing, so do others. As they gain his information, they may gain more at his expense than he hopes to gain from their information.

information sharing, so that the assumption of actual information sharing follows.

Some social institutions promote information sharing through compensation – for example, the practices of patents, copyrights, and royalties. Obstacles to information sharing in real-life group decisions arise from difficulties in providing adequate payment for information shared. In the ideal case, I suppose that all obstacles to information sharing are overcome: compensation for sharing information is possible, mechanisms for compensation are available, and rational agents utilize them.

Under the idealization, compensation prevents losses from sharing information. To make the process more vivid, I apply the idealization to bargaining problems concerning divisions of commodities for which markets and trading possibilities exist. I assume that the value of information in these bargaining problems is given by the increase in the amount of a commodity the information brings its possessor. To make information sharing rational, I suppose an institution exists for compensating the sharer for the value of the information shared.

How is compensation determined? Is there a marketplace for information that sets the value of information, so that a group of bargainers purchases the private information of members at its market value? No, the market for information does not always give its value to the holder. For instance, because the buyer of information does not know what he is getting, the seller may have to provide a discount to make a sale. I imagine that an arbitrator sets the compensation for revelation of private information. First, the arbitrator uses general bargaining theory to calculate the outcome (in terms of commodities) that would result from bargaining without sharing information. Then the arbitrator calculates the outcome (in terms of commodities) that would result if information were shared. These outcomes may differ because individual utility assignments change with the acquisition of new information. Suppose that the outcomes differ. A change from the outcome without information sharing to the outcome with information sharing is advantageous to some individuals and disadvantageous to others. Some individuals have an incentive to resist information sharing. Given the idealizations, however, those gaining from information sharing may compensate those losing for the market value of their commodity losses. With compensation, sharing relevant information is Pareto-optimal and rational for all.

194

To illustrate, suppose that a buyer is ready to pay $1 million for some land and hopes to sell oil he knows is there for $1 million and then resell the land for $1 million netting a profit of $1 million. Hence his knowledge about the oil has a value for him of $1 million. After information sharing, the seller may keep the land and sell the oil himself for $1 million. That million may be transferred to the buyer as compensation for the information shared. Similarly, the owner of a defective car may be able to sell it for $1,000 if information about the defect is his alone, and he may be able to sell it for only $100 if the information is shared. Then the value for him of his information is $900. He can be compensated for sharing his information. After information sharing, the prospective buyer may refuse the car and save $900, the difference between the sale price and the informed market value of the car. The savings can be used to compensate the seller for sharing information. An arbitrator may supervise compensation.

Compensation is not always possible. Knowledge of the meaning of ancient secret signs may have value in a community only if it is exclusive knowledge, a matter of privilege. If shared, the information may have no value. Then information sharing cannot be compensated. My idealization puts aside such cases. I treat cases, such as selling the land and the car, where information sharing produces gains that can be used to compensate losses. Information sharers receive their information's value to them. They suffer no net loss after compensation supervised by an arbitrator.

I claim that it is a goal of rationality for a group to maximize expected group utility using the common probability assignment that results from information sharing. The group should strive to achieve this goal even in the absence of a common probability assignment. It is a goal of rationality for a group, like the goal of Pareto optimality. When obstacles are removed, the goal's attainment is the result of the group's members individually attaining the goals of rationality for themselves.[16]

Given that maximizing expected group utility with a common group probability assignment is a goal of rationality for a group, conditions

---

[16] In nonideal conditions the group probability assignment is a hypothetical probability assignment, the one that all would adopt after information sharing. The actual group probability assignment is the hypothetical common probability assignment. Having that common probability assignment is a goal of rationality for the members, and that assignment is used to formulate for the group a goal of rationality – maximizing expected group utility.

conducive to the formation of that common probability assignment are included in the conditions ideal for group action. The common probability assignment is, in effect, included in the idealization that conditions are perfect for group action. It does not follow from ideal conditions for group action, without the rationality of the members, that all have the same probability assignment. Those ideal conditions just provide the means of meeting the goals of rationality. Given the rationality of the members as part of the overall idealization, however, a common probability assignment does follow. In ideal conditions for group action, where there are incentives to share information, rational agents pool their information. Given the additional explanatory idealization about the pooled information's richness, it legislates a unique probability assignment. Hence, with all idealizations in force, the members of the group have the same probability assignment.

The idealization about information sharing is strong. It puts aside a significant influence on group action, namely, motivated concealment of information. However, the idealization is justified, as Section 1.3 explains, because it throws into relief other, more tractable explanatory factors. The theory it generates treats cases in which rational group action springs from optimal use of a group's resources, including information. These are theoretically important cases. They are especially important in the next chapter, which addresses trustee decisions. Often an expert deciding for a group has all the relevant information the group's members have. To serve the group well, she may consider what it would do if its members had her information. Thus she imagines a case in which they all have the same information. She explores a case arising under the idealization of shared information. The theory the idealization generates thus aids trustee decision makers.

6.4.2. Intrinsic and Group Utility Analyses

Next, I apply intrinsic utility analysis to group utility. Groups do not have basic intrinsic desires in a literal sense. So to apply intrinsic utility analysis, one must define basic intrinsic desires for a group. I take these as the basic intrinsic desires of the group's members. Every basic intrinsic desire of a member yields a basic intrinsic desire of the group. Basic intrinsic aversions for a group are introduced in a similar way. These basic intrinsic attitudes (BITs) of a group may not meet all the coherence requirements for individuals. Nonetheless, they may be the components of an intrinsic utility analysis of the group's utilities because group utilities resolve all conflicts between the members' desires. Another component of the analysis is group intrinsic utility GIU, which I define following the pattern of the definition of group utility, after putting members' intrinsic utilities on the same scale:

$$GIU(p) = \Sigma_i e_i IU_i(p).$$

When applied to the object of a BIT for a group, the formula yields the attitude's intensity for the group. For each member with the basic intrinsic attitude, the formula multiplies the economic power of the member and the interpersonal intensity of her attitude, and then adds the products.

Intrinsic utility analysis for a group applies the general form of intrinsic utility analysis, the form that handles uncertainty, to $GU(o)$ using GIU's of objects of the group's BITs. In the formula for $GU(o)$, I let BIT_j stand for the object of the basic intrinsic attitude BIT_j and let GP stand for the group's probability assignment after pooling information. The result is:

$$GU(o) = \Sigma_j GP(BIT_j \text{ given } o)GIU(BIT_j),$$

where BIT_j ranges over the objects of basic intrinsic group attitudes.

To extend intrinsic utility analysis to conditional group utilities, I apply the foregoing formula under the supposition that a state s holds. This yields the following formula:

$$GU(o \text{ given } s) = \Sigma_j GP_o (BIT_j \text{ given } s)GIU(BIT_j).$$

In the group probabilities o and s are supposed differently and so separately. The option o is supposed subjunctively and the state s is supposed indicatively. As Section 4.2 explains, the appropriate form of supposition for a condition is subjunctive if the condition is an option and indicative if it is a state.

6.4.3. Consistency

Given that expected and intrinsic utility analyses apply to group utility, the group utility of an option may be calculated several ways. For example, one may apply group utility analysis to possible outcomes of the option and then use expected utility analysis to obtain the utility of the option, or one may use expected utility analysis to obtain the utility for individuals of the option and then use group utility analysis to obtain the group utility of the option. Intrinsic utility analysis multiplies the methods of computing group utility by adding another dimension of analysis. Are all these analyses consistent, at least given ideal conditions for rational group action?

The various ways of computing $GU(o)$ in fact all assign the same value to it. My methods of analysis, which apply consistently to utility for individuals, also apply consistently to group utility. The computations of $GU(o)$ they yield are consistent because expected utility analysis and intrinsic utility analysis both follow from the principle of pros and cons. All computations of $GU(o)$ partition the same considerations, and any partition of the same considerations yields the same sum of pros and cons. As Section A.6 shows, the various ways of computing $GU(o)$ reduce to sums of the same probabilities of realizations of BITs, just organized in different ways.

6.5. BARGAINING IN A POWER STRUCTURE

In support of my definition of group utility, I have claimed that a rational choice by a group, defined as a product of rational choices by its members, maximizes group utility, defined as a power-weighted sum of members' utilities, when conditions are ideal for group action and group utilities exist. I call this position *weighted utilitarianism*. Although the principle of pros and cons is the main motivation for my definition of group utility, this section provides additional support for the definition by strengthening the case for weighted utilitarianism.

To argue for weighted utilitarianism, I show that in certain bargaining problems, given all the information its application requires, weighted utilitarianism clearly yields the rational group choice. The demonstration uses only the noncontroversial standard of Pareto optimality as the criterion of rational group choice.[17] It shows that in the bargaining problems treated, Pareto optimality yields the weighted

---

[17] Broome (1991: chap. 7) objects to the Pareto principle applied to preferences, its usual application, and proposes instead the principle of personal good (chap. 8). His objections concern the nonideal case where agents are irrational. Indeed, Pareto optimality is not a plausible standard given irrational preferences. In that case, objective betterness is more reliable than personal preferences. However, my application of Pareto optimality to utilities avoids Broome's criticism because it is limited to cases with fully rational agents. Given the idealizations, preferences and degrees of desire are fully rational and conform with objective betterness and goodness, respectively, if we assume that rationality requires conformity. The principle of Pareto optimality is sound for rational ideal agents.

utilitarian solution.[18] Although the demonstration considers weighted utilitarianism only in bargaining problems meeting certain simplifying restrictions, weighted utilitarianism's success in these restricted problems makes its general success more plausible. I leave for future work the task of providing supplementary general arguments that support weighted utilitarianism without reliance on restrictions.

Adopting this chapter's background idealizations, I treat bargaining problems with ideal agents in an ideal situation for group decisions, except perhaps for uncertainty. In the context of bargaining problems I take the idealizations to include agents' common knowledge of their rationality because solutions depend on the rationality of all agents.[19] Besides these idealizations, I impose some restrictions. I suppose that bargaining takes place within a power structure so that the participants have a means of obtaining bargaining concessions by offering rewards for them or by imposing penalties if concessions are not granted. I also suppose that power consists specifically in an ability to transfer utility to individuals in order to obtain concessions. These suppositions ensure that weighted utilitarianism applies in a straightforward way. In the overall game in which the bargaining problem is embedded, quantitative power is harder to assess and weighted utilitarianism harder to apply.

In the group decision problems to which I apply weighted utilitarianism, the options are various resolutions of a bargaining problem embedded in a power structure. The options do not themselves include the exercise of power by individuals. To avoid complications, I also take the power structure to be static. That is, bargaining outcomes have a negligible effect on the power structure. Agents cannot exchange part of their gains under an outcome for a concession from an opponent – potential payoffs cannot be parlayed into power. Agents can use only the background power structure to purchase concessions. Furthermore, the purchase of a concession does not affect the power structure. The purchase prices are negligible with respect to the power structure, or have a delayed effect on it.

Given the restrictions, the power structure and interpersonal utilities are used to define group utilities for the outcomes of the embedded

---

[18] Some technical apparatus for my argument is from the theory of cooperative games with transferable utility, as presented, for example, by Shubik (1982: chaps. 6, 7). This theory does not assume that utility is interpersonal, but my argument does. The assumption grounds my account of power.

[19] Common knowledge here is public knowledge – that is, all know that all are rational, know that all know that all are rational, and so on.

bargaining problem. Weighted utilitarianism is then supported by showing that in these restricted bargaining problems the rational group action is in fact the weighted utilitarian outcome. I now begin a more detailed presentation of the bargaining problems treated.

The members of a group of people can often cooperate in ways that benefit all. For cooperation to occur, however, they typically must agree on a division of the benefits of cooperation. In such cases I say that the group has a bargaining problem. For example, a will may stipulate that the deceased's family must agree on a division of the estate in order to inherit any of it. Then the family faces a bargaining problem.

The classical approach to a bargaining problem, which Section 6.A.2 briefly reviews, begins with a representation of the problem in utility space. A typical problem for two people might have the representation in Figure 6.4. S stands for the set of possible outcomes, and d stands for the outcome if bargaining breaks down – that is, reaches the disagreement point. For simplicity, I adopt the convention of setting the zero point for utility gains so that it represents the disagreement point. Also, I represent only possible outcomes that meet the standard of individual rationality, and hence offer each agent no less than the disagreement point, which any agent can realize by unilateral action. Although the utilities of outcomes for individuals are represented together, they are not interpersonally comparable in the classical representation of a bargaining problem.

I enrich the classical representation by adding interpersonal comparisons of utility and assessments of the economic power of the bargainers. These additional features make it easier to justify solutions. The representation of interpersonal utilities is an easy addition. It requires only that the unit of utility be the same for each individual. The representation of power, on the other hand, requires some new technical apparatus. As mentioned, I take power as the ability to provide rewards or impose losses. I assume that power is constant in a particular problem, and that rewards are incentives just as good as penalties so that power is exercised exclusively through rewards, or side payments.[20] Given these assumptions, power relations may be repre-

---

[20] Providing incentives through threats may be more effective than providing incentives through rewards if the commodity used to transfer utility has diminishing marginal utility. Also, the advantage of providing incentives through (binding) threats is that if successful one obtains what one wants at no cost to oneself. The risk is that the threat fails and one incurs a cost without any compensating gain. Although the cost of making the threat itself is zero in ideal conditions, the cost of imposing

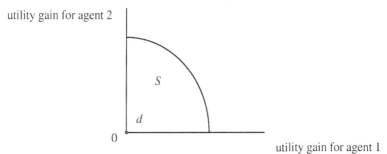

utility gain for agent 2

S

d

0

utility gain for agent 1

Figure 6.4. A Classical Bargaining Problem

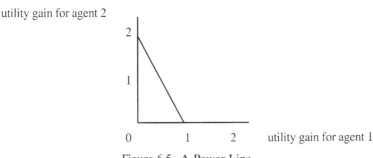

utility gain for agent 2

2

1

0   1   2   utility gain for agent 1

Figure 6.5. A Power Line

sented by lines in utility space indicating possible transfers of utility from one person to another. Suppose that by giving up one unit of utility, agent 1 can produce a two-unit gain for agent 2. Then I say that agent 1 is twice as powerful as agent 2. Figure 6.5 represents the power of agent 1 with respect to agent 2 using a line of slope −2.

Given interpersonal utilities and power relations, the solution to a bargaining problem is clear. In a typical problem there is only one outcome that is Pareto-optimal in light of possibilities for side payments.[21] That is, only one outcome is such that there are no alter-

the threatened penalty is not zero. However, I suppose that ideal bargainers consider the results of possible threats. They work their way toward a solution acceding to threats, which are not made unless accession is predicted, so that the cost of carrying out a threat, or imposing a penalty, is not incurred. Nonetheless, no one gains an advantage if all are free to use threats. So I assume, for simplicity, that all use rewards. This will not skew the solution in anyone's favor.

[21] In an atypical problem I take the solution to be the set of outcomes that are Pareto-optimal given the power structure; every outcome in this set is a rational outcome.

natives that are Pareto-superior, that is, no alternatives in which some agents do better and none do worse in terms of their personal utility assignments. Clearly, rational agents would agree upon that outcome in ideal conditions because such conditions remove obstacles to negotiations.[22]

In a two-person bargaining problem, the Pareto-optimal outcome is the intersection of S with the highest intersecting power line. Every other bargaining outcome is dominated by outcomes reached by moving to the point of intersection, u, and having the gainer compensate the loser. In Figure 6.6, for example, agent 1 loses in the move from u' to u, but agent 2 can compensate him by transferring utility so that after side payments the net result is u'', a result better than u' for both agents. Notice that as the power of agent 1 increases, the power line becomes steeper, and the intersection point moves along the boundary of the bargaining problem closer to the horizontal axis so that the solution to the problem is more favorable to agent 1, exactly as one would expect. More generally, in an n-person bargaining problem, the Pareto-optimal outcome is the intersection of S with the highest intersecting hyperplane whose slope with respect to the axes for a pair of agents is the slope of a power line for that pair.[23]

To see that the intersection point is the weighted utilitarian outcome, consider the following. In the two-person case, the negative of the slope of a power line is the ratio of agent 1's economic power to agent 2's, that is, e_1/e_2. Hence the equation of a power line is $x_2 =$

The outcome realized may depend on the protocol for bargaining, that is, on information about the bargaining problem beyond this section's scope.

[22] Coase (1960) forcefully argues that rational bargainers settle on a Pareto-optimal agreement in the absence of transaction costs. His conclusion is known as Coase's theorem. In ideal conditions for group action, there are no transaction costs. In accordance with Coase's theorem, I then assume that a rational group action is Pareto-optimal.

[23] In the n-person case I represent the power relations, or the rates of utility transfer, by an ordered n-tuple $(e_1, e_2, \ldots, e_n)$, where e_i gives the power of agent i. For individuals i and j the number of units of utility from i required to produce one unit of utility for j is e_j/e_i. This method of representation assumes that the rates are related in various ways. For example, the rate of transfer from j to i must be the reciprocal of the rate of transfer from i to j. Also, if the rate of transfer from i to j is r_1, and the rate of transfer from j to k is r_2, then the rate of transfer from i to k must be $r_1 r_2$. To ground these constraints, one may suppose that utility transfers are made with a resource that has linear utility for each individual. Given this representation of power relations, the solution to an n-person bargaining problem is the intersection of S with the highest intersecting hyperplane whose slope with respect to the i and j axes is $-e_i/e_j$.

utility gain for agent 2

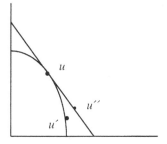

utility gain for agent 1

Figure 6.6. Dominance via the Intersection Point

$(-e_1/e_2)x_1 + k$, or $e_1x_1 + e_2x_2 = e_2k$, where e_2k is a constant. That is, a power line is a locus of points having the same power-weighted sum of utilities. Because e_2k is at a maximum for the highest intersecting power line, u maximizes the power-weighted sum of interpersonal utilities. In the n-person case, one may obtain the same result in an analogous way. Hence in bargaining problems embedded in a static power structure, maximizing group utility in compliance with weighted utilitarianism is equivalent to achieving Pareto optimality. Section 6.A.3 supports the argument by considering the role of coalitions in embedded bargaining problems. Section 6.A.4 supports it by considering the reduction of embedded bargaining problems to a series of individual choices.[24]

To avoid controversy, I use only embedded bargaining problems to confirm weighted utilitarianism. However, solutions to embedded bargaining problems are not solutions to the overall games in which bargaining is embedded. They do not detail side payments. Although group

[24] Weighted utilitarianism resembles proposals for bargaining solutions that introduce measures of bargaining power. For example, it resembles the asymmetric Nash solution – in particular, versions in which risk-sensitivity supplies weights as in Roth (1979: pt. I) or in which time-discounting supplies weights as in Rubinstein (1982) – although these proposals do not use interpersonal comparisons of utility. It also resembles Shapley's (1969) proposed solution to cooperative games with transferable utility, although Shapley defines interpersonal utility in terms of a game's solution and supports his proposal with standards of equity in addition to standards of rationality. Harsanyi (1955) proposes solving social choice problems by comparing options according to sums of weighted utilities, as weighted utilitarianism does, although the utilities he uses are not interpersonal and the weights do not measure economic power. Also, general equilibrium theory's method of resolving trading problems is similar in spirit to weighted utilitarianism. Initial endowments and prices in general equilibrium theory have roughly the same function as power structures and interpersonal utilities in weighted utilitarianism.

203

decision making takes place over time and in a social context, so that the solutions to embedded problems are significant even if not final results, and although embedded problems must be solved en route to final solutions of the overall games, an important additional test of weighted utilitarianism is whether it yields solutions to the overall games, not just embedded bargaining problems. To reach solutions to the overall games, one needs principles more controversial than Pareto optimality. Weirich (1993) extends the argument for weighted utilitarianism to the overall games, using a principle of symmetry to argue that a solution of an embedded bargaining problem is also a solution of the overall game. It argues that side payments are not needed to obtain a solution to the overall game; the possibility of them suffices. It concludes that the solution to the overall game is its weighted utilitarian solution.

This argument treats the overall n-person cooperative game for an embedded bargaining problem, but not every n-person cooperative game. One difficulty in extending the argument to all cooperative games is formulating a detailed account of power in such games. In embedded bargaining problems power reduces to ability to make utility transfers. In a bargaining problem of another type, it may depend on the shape of the problem's representation in utility space, the agents' abilities to sustain bargaining failure, the protocol for bargaining, the agents' urgency for settling, and the like.

6.A. APPENDIX: WEIGHTED UTILITARIANISM, SOCIAL CHOICE, AND GAMES

This appendix defends my position about rational group choice, weighted utilitarianism, in light of some well-known claims about rational group choice. It reviews social choice theory and bargaining theory, and enriches Section 6.5's argument for weighted utilitarianism's application to embedded bargaining problems.

Weighted utilitarianism bears on social choice theory and bargaining theory. It needs to be assessed with respect to the work of Arrow (1951) and Nash (1950), work that may seem to undermine weighted utilitarianism's account of rational group action. I argue that weighted utilitarianism survives confrontation with that work because it approaches group decision problems with greater informational resources than that work assumes. Unlike classical social choice and bargaining theory, weighted utilitarianism uses interpersonal comparisons of utility. Unlike classical bargaining theory, weighted utilitarianism uses measures of bargaining power sensitive to factors besides the shape of a bargaining problem's representation in utility space. Weighted utilitarianism does not contest the results of Arrow and Nash when informational resources are limited. When informational resources are expanded, it is unconstrained by their results.

6.A.1. Social Choice

Social choice theory seeks a definition of group preferences. A famous theorem by Arrow (1951: chap. 5) shows that it is impossible to define group preferences in terms of individual preferences. Arrow's theorem shows that aggregating individuals' preference rankings into a group's preference ranking inevitably violates certain plausible conditions, such as Pareto optimality.[25]

The theorem has momentous consequences for voting mechanisms whose input is restricted to individuals' preference rankings, and for nonideal cases in which a definition of group preferences must use only individuals' preference rankings. But suppose resources go beyond individuals' preference rankings. Then various promising ways of addressing social choice appear, as Binmore (1994: 132–3) notes. For instance, Arrow's theorem does not refute utilitarian principles of social choice. An appeal to interpersonal comparisons of utility in addition to individuals' preferences provides a means of circumventing Arrow's problem, as shown, for example, by Strasnick (1975), Hammond (1976), D'Aspremont and Gevers (1977), and Roberts (1980). In particular, the standard utilitarian definition of group preferences in terms of interpersonal utilities meets Arrow's conditions, once reformulated to accommodate interpersonal utilities.

Weighted utilitarianism uses interpersonal utilities and power distributions in defining group utility. It suggests a way of defining group preferences in terms of group utilities: if one option has a greater group utility than another, the group prefers it to the other. This definition also meets appropriate analogues of Arrow's conditions. Because it

---

[25] For discussion of the theorem, see Sen (1970), MacKay (1980), and Suzumura (1983).

205

uses more than individuals' preference rankings, it does not run afoul of Arrow's theorem.[26]

Contextualism about theoretical concepts provides welcome resources for social choice theory. Using interpersonal utilities, a contextualist may define group preferences, and extend decision principles concerning preferences to groups of individuals and their actions, at least in ideal cases. On the other hand, an operationist must either abandon the project of defining group preferences, or else settle for a definition that compromises the relevant desiderata.

6.A.2. Bargaining

This section defends weighted utilitarianism's application to a certain type of group decision problem, namely, bargaining problems. These problems belong to a branch of game theory reviewed, for example, by Harsanyi (1977: pt. 3), Roth (1979, 1985), Binmore and Dasgupta (1987), Elster (1989: chap. 2), and Binmore (1998: chap. 2, app. C).

Theorists such Nash (1950), Kalai and Smorodinsky (1975), and Gauthier (1986: chap. 5) advance various proposals about solutions to bargaining problems, or rational group action in bargaining problems.[27] Their proposals may seem to be at variance with weighted utilitarianism. But they do not address the same problems that it addresses. In the bargaining problems they consider there is information about personal utility assignments, but no information about interpersonal comparisons of utility, nor about economic power except what is derivable from a bargaining problem's representation in utility space.[28] In other words, they attempt to solve bargaining problems less rich in information than the problems that I try to solve.[29] The solutions they propose are unin-

---

[26] For weighted utilitarianism, the analogue of Arrow's independence condition says that a group preference ranking of x and y depends only on the members' utility assignments to x and to y. The analogue of Arrow's nondictatorship condition says that no single member's preference ranking is decisive in all cases unless he has all the economic power. For additional discussion of various approaches to social choice, see Weirich (1988a).

[27] They avoid Arrow's impossibility results in part by treating only bargaining problems with the property of convexity and so by abandoning the principle of universal domain.

[28] Brandenburger and Nalebuff (1996: 40) explain classical game theory's conception of power this way: "Power – yours and others' – is determined by the *structure* of the game. Game theory shows how to quantify this power."

[29] The standard approach considers only two properties of a bargaining problem, namely, the representation of the set of alternatives in utility space, S, and the

tuitive, given the richer informational resources of the problems I consider. In these problems exogenous power relations between the bargainers swamp endogenous bargaining advantages. The difference in informational resources explains the difference in solutions proposed by classical bargaining theory and weighted utilitarianism.

Of course, studying solutions to bargaining problems where informational resources are restricted is worthwhile. Bargaining behind a veil of ignorance is a cornerstone of contractarian moral theory, for example. However, weighted utilitarianism applies only to bargaining problems where informational resources encompass interpersonal utilities and power. So I restrict myself to these problems. Given my idealizations, Section 6.5 shows that rational bargaining in these problems conforms with weighted utilitarianism.

In classical game theory bargaining problems are abstract representations of concrete situations. However, I take bargaining problems as concrete situations and therefore as having more features than their representations depict. Classical bargaining problems sometimes do not have enough information to reveal solutions to the concrete bargaining problems they represent. The usual representation of a bargaining problem does not provide full information about power. The representation's incompleteness obscures the solution of the concrete bargaining problem. The solution of the concrete problem depends on the facts and not merely the information provided by its representation. In a theory of rationality it is often appropriate to take bargaining problems as abstract representations because the rational course of action depends on the information available to agents rather than on the facts, which may be unknown. Under this chapter's idealizations, however, agents know all the relevant facts about their bargaining problem. They are fully informed about their game and each other. So I look for solutions to concrete bargaining problems, not abstract representations of them. An abstract representation may serve as a

disagreement point, d. The Nash (1950) and Kalai-Smorodinsky (1975) solutions to bargaining problems, for example, are functions of (S, d). The parsimony of the standard approach makes the solutions advanced very controversial. They try to get much out of little. In particular, the principles used to support the solutions are controversial. The principle of independence used to support the Nash solution and the principle of individual monotonicity used to support the Kalai-Smorodinsky solution are objections to each other. Luce and Raiffa (1957: 128–34) present objections to the Nash solution and, in particular, to the principle of independence (pp. 132–3).

substitute for a concrete problem only if it contains all the relevant facts, that is, all the information needed to find a solution.

According to a common objection, the extra informational resources that weighted utilitarianism uses to obtain solutions to bargaining problems are superfluous; interpersonal utility has no influence on rational group actions because rational agents act according to their personal utility functions. Appearances notwithstanding, interpersonal utilities do affect the behavior of rational individuals in bargaining problems. The objection overlooks the influence of interpersonal utilities on personal utilities via beliefs about interpersonal utilities. Astute agents perceive interpersonal utilities. Their beliefs influence the utilities they assign to options in bargaining problems and hence affect the rational course of action.

For instance, a stall owner at a bazaar lowers his price on an item as a haggler walks away depending on whether he thinks the haggler wants the item a little or a lot. The stall owner's action depends on his beliefs about interpersonal comparisons of utility. To decide whether to lower his price, he compares the haggler's intensity of desire to the expected intensity of desire of future clients to whom he might sell the item. His offer depends on his estimate of the haggler's relation to others who are in the market for the item.

He does not lower his price if he expects a future client willing to pay that price. His expectations about future clients depend in part on his assessment of the haggler's desire for the item. Is it weak or strong in comparison with prospective clients' desires for the item? If it is weak, he may wait for a new client, hoping to find one willing to pay more than the haggler offers. If it is strong, he may lower his price, thinking the haggler is his best chance to make a sale. Although the stall owner may dispense with interpersonal comparisons of desires, given knowledge of the prices the haggler and prospective clients are willing to pay, being ignorant of those prices, his best way of estimating them may require interpersonal comparisons of desires. Clients may reveal comparative strengths of desire, although other factors relevant to willingness to pay, such as wealth, are not apparent.[30]

---

[30] In certain situations Binmore (1998) advances a bargaining solution resembling power-weighted utilitarianism. He argues that in a recurrent two-person bargaining problem the bargainers empathetic preferences will eventually agree as a result of social evolution (p. 178). The utilitarian solution with personal utilities weighted by their worth according to those empathetic preferences agrees with the bargaining problem's Nash solution, and hence those weights indicate bargaining

6.A.3. Coalitions

Let us consider an objection to weighted utilitarianism as it applies to embedded bargaining problems, the problems Section 6.5 treats. The objection arises in n-person bargaining problems where a splinter group or coalition can obtain benefits for its members without the cooperation of others. It might turn out that the coalition can do better on its own than under some Pareto-optimal outcome. Then it has an incentive to block that outcome by withholding consent. In fact, it is possible that for each Pareto-optimal outcome some coalition has an incentive to block that outcome. In this case the *core* of the problem, the set of Pareto-optimal outcomes that no coalition has an incentive to block, is empty. It appears that no Pareto-optimal outcome is an equilibrium, in particular, not the Pareto-optimal outcome provided by weighted utilitarianism. Because being an equilibrium is a prerequisite for a rational group action, it then seems that weighted utilitarianism does not yield a rational group action.[31]

To illustrate, let v designate the *characteristic function* for a bargaining problem concerning the division of a certain commodity. The function v yields the *value* of each coalition, or the amount of the commodity that each coalition can obtain on its own. I assume that this amount may be divided among the coalition's members as they please. Suppose that the bargaining problem involves three individuals A, B, and C. Also, suppose that the values of the possible coalitions are as follows:

$$v(A) = v(B) = v(C) = 0.$$
$$v(AB) = v(BC) = v(AC) = 8.$$
$$v(ABC) = 9.$$

power (pp. 255–6, 449–50, 456–8). Binmore's power-weighted utilitarianism differs from mine in significant ways, however. First, I use interpersonal utilities rather than personal utilities. Second, my weights indicate power, not worth, and even where worth and power in Binmore's senses agree, my weights indicate power defined in terms of the ability to transfer utility and not power defined in terms of the bargaining problem's Nash solution. In contrast with Binmore, I do not define interpersonal utilities in terms of empathetic preferences, nor do I define power in terms of Nash's solution to a bargaining problem. In accordance with contextualism, I use interpersonal utilities and power to explain solutions to bargaining problems. Because of these differences, my version of power-weighted utilitarianism may disagree with Binmore's version (and with Nash's solution) even in a recurrent bargaining problem after social evolution has had time to operate.

[31] For a discussion of equilibrium in noncooperative games, see Weirich (1998).

No distribution of the commodity is in the core because none gives each coalition at least as much as it can obtain on its own. The most available for distribution is 9, and 9 cannot be divided so that each pair of individuals receives 8 or more. The distribution $(3, 3, 3)$, for example, does not give any pair 8 or more. So each pair has an incentive to block it. For instance, A and B can form a coalition that brings each more than 3. They can achieve the distribution $(4, 4, 0)$. Similarly, for every Pareto-optimal distribution, some pair has an incentive to block it.

The reasons for a failure to achieve Pareto optimality in bargaining problems are different from the reasons for a failure to achieve Pareto optimality in the Prisoner's Dilemma (see Section 6.2). In the Prisoner's Dilemma failure arises because of nonideal conditions, namely, the absence of communication and the means of making binding contracts. In the bargaining problem discussed, a failure to achieve Pareto optimality, if it occurs, occurs because of the structure of the possible outcomes of the problem. It occurs despite ideal conditions for cooperative games.

I argue that in bargaining problems embedded in a power structure, rational agents in ideal conditions achieve Pareto optimality in spite of incentives for coalition formation. In these bargaining problems power is exercised through side payments to bind all coalitions to a Pareto-optimal agreement. For example, suppose that three agents hoping to collaborate in a moneymaking enterprise must first agree on a division of the profits. Suppose that money has diminishing marginal utility for them so that the more money they receive the less utility they derive from an additional dollar. Figure 6.7 represents the possible outcomes of their bargaining problem in terms of utility. Utilities for agents $1, 2$, and 3 are along the x, y, and z axes, respectively. The plane indicates power relations. The weighted utilitarian outcome, the intersection point u, represents a division of profits in terms of utility, specifically, $(4, 1, 1)$. Possible transfers of utility given u stay in the plane. The point u' stands for a division of profits from collaboration between agents 2 and 3 only, namely, $(0, 2, 2)$. Although agents 2 and 3 benefit from the move from u to u', the profits from their exclusive collaboration are less than the profits from the collaboration of all three agents. Hence it behooves agents 2 and 3 to collaborate with agent 1 to achieve u with the understanding that he then transfers utility to realize, for example, u'', that is, $(1, 3, 3)$. This arrangement makes all three agents better off than they are with u'.

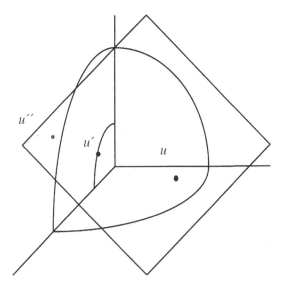

Figure 6.7. Stability of the Intersection Point

In general, the intersection point u is a stable outcome of the bargaining problem even taking account of coalitions' incentives. It is a stable outcome of the embedded bargaining problem because side payments, if necessary, prevent coalitions' defection. For suppose that some coalition has an incentive to block u because another outcome u' is better for all its members. Given the possibilities for concession and compensation that the problem's power structure allows, another outcome u'' can be reached through u and is better than u' for everyone. Hence the entire group of bargainers has an incentive to block u' and adopt u. Because the weighted utilitarian outcome of an embedded bargaining problem is an equilibrium given the power structure, I conclude that it is indeed a rational outcome of the bargaining problem. Hence my investigation of bargaining problems supports weighted utilitarianism, even in cases where coalitions may form.

6.A.4. Noncooperative Games

Section 6.5 takes embedded bargaining problems as cooperative games, where the group's decision might just as well have been reached by an arbitrator seeking the group's welfare, and the support for a solution depends on standards for rational joint action, such as Pareto

211

optimality. It assumes that rational bargainers devise a way to reach a Pareto-optimal outcome, given ideal conditions for group action, including opportunities for communication and binding contracts. Then it shows that weighted utilitarianism yields a cooperative solution to bargaining problems. That is, it yields a solution, given the assumption that bargainers work out their differences to obtain the benefits of cooperation. The ultimate demonstration of the rationality of a group action, however, is showing that the action results from rational actions of the members. So now I consider embedded bargaining problems as noncooperative games where the individuals bargain for themselves, and the outcome depends on the way in which it is rational for them to bargain. Actions are unilateral, not joint, and not actions of an arbitrator for the group. I want to show that weighted utilitarianism also yields a noncooperative solution to embedded bargaining problems in ideal conditions for group action. That is, I want to show that rationality leads to the weighted utilitarian outcome without taking it for granted that some form of cooperation occurs. This section therefore puts aside Pareto optimality as a basic standard of rational group action and uses only standards of individual rationality to support the weighted utilitarian outcome of an embedded bargaining problem.[32]

My project relies on a method of reducing a cooperative game to a noncooperative game. For simplicity, I consider only embedded bargaining problems. Taking them as noncooperative games requires saying more about bargaining procedures because those procedures affect the strategies bargainers follow. I take offers, acceptances, payments, threats, promises, and concessions as unilateral actions open to the bargainers. A full protocol specifies who may exercise individual options such as making an offer, when they may exercise those options, what coalitions are possible, and anything else that influences rational

---

[32] Deriving cooperative solutions from more fundamental noncooperative solutions is called the Nash program, after its earliest advocate, John Nash. The Nash program views solutions as negotiated, not arbitrated. Rubinstein (1982) carries out the Nash program for a case in which two bargainers make offers alternately. He shows that a cooperative solution of their bargaining problem, the generalized Nash solution, is a noncooperative solution of the game of making offers alternately, a subgame-perfect equilibrium. Binmore (1998: 171) shows that, given some assumptions about the bargaining problem, Rubinstein's results agree with a modification of weighted utilitarianism. This section sketches another route by which noncooperative considerations bolster weighted utilitarianism.

strategies. I ask what outcome the bargainers will achieve using their individual options.[33]

This section only illustrates the results I hope to establish for non-cooperative embedded bargaining problems. It addresses only the following two-person case. Imagine that agents 1 and 2 have to agree upon a division of D dollars between themselves. Money has diminishing marginal utility for each. They bargain by letting one make a proposal and the other either accept it or make a counterproposal. Each may make binding promises to pay a reward in return for a concession. What division will they agree upon if they are rational?

Informally, think of the problem as follows. The possible divisions of money are represented in terms of utility by a curve like the one in Figure 6.8. In the figure, u represents a proposed division. As u moves along the curve closer to the division that is optimal for agent 1, the division represented by the intersection of the curve with the horizontal axis, agent 1 gains less and less utility and agent 2 loses more and more utility. So agent 2 has a larger and larger incentive to reward agent 1 for not pursuing further gains, and agent 1 has less and less incentive to spurn his reward. Conversely, as u moves along the curve closer to the division that is optimal for agent 2, the division represented by the intersection of the curve with the vertical axis, agent 1 has a larger and larger incentive to reward agent 2 for not pursuing

---

[33] An extremely restricted protocol for individual action is to let arbitration be followed by opportunities for destabilizing acts. First, an arbitrator imposes a resolution of the bargaining problem. Then, individuals have an opportunity to upset it by trading resources for concessions in the bargaining problem. Considering stability after arbitration introduces noncooperative considerations, noncooperative because a matter of agents' individual incentives. The question is whether any bargainer has reason to perform some unilateral option that would upset the arbitrated solution. Maximizing group utility passes the noncooperative test given the protocol. It is stable. No one can do better by pushing for a different resolution of the bargaining problem. An arbitrated solution that maximizes group utility achieves Pareto optimality in the net outcome after side payments. Hence weighted utilitarianism is a solution to a noncooperative game with this protocol.

It may turn out that many protocols are equivalent in the sense of leading to the same solution. A more detailed investigation might characterize the family of *sensible* protocols for bargaining. A partial characterization might stipulate, for example, that if P is a sensible protocol according to which agent 1 starts, then if P' is a protocol just like P except that according to it agent 2 starts, then P' is also a sensible protocol. After achieving a full characterization of sensible protocols, the investigation might show that every sensible protocol yields the weighted utilitarian solution.

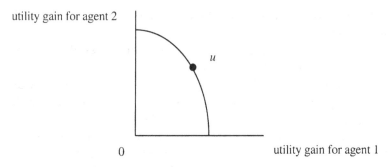

utility gain for agent 2

u

0   utility gain for agent 1

Figure 6.8. Divisions of *D* Dollars

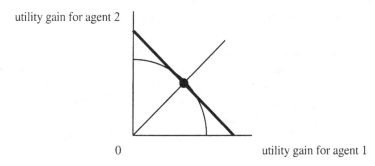

utility gain for agent 2

0   utility gain for agent 1

Figure 6.9. Equilibrium of Incentives

further gains, and agent 2 has less and less incentive to spurn his reward. It is as if agents 1 and 2 were in a tug of war where the more ground one loses the harder he pulls and the less hard his opponent pulls. As a result, there is a division that constitutes an equilibrium in their "tug of war." That equilibrium division is the solution to their problem. It identifies a proposal that elicits acceptance instead of an offer of a reward for a further concession.

For example, suppose that agents 1 and 2 have equally strong desires for money so that the representation of possible divisions of money in terms of utility is a curve that is symmetrical about a line through the origin, making a forty-five degree angle with each axis. Also suppose that agents 1 and 2 have the same amount of power so that a power line representing the power of agent 1 over agent 2 has a slope of −1. Then the equilibrium division is the division represented by the point where the slope of the tangent to the curve equals −1, namely, the equal division. See Figure 6.9.

214

In general, equilibrium occurs at the point on the curve where the slope of the tangent line equals the slope of a line representing the power of agent 1 over agent 2. Equilibrium occurs at that point because the slope of a power line measures the pressure that agent 1 can exert on agent 2, and the slope of the tangent line measures agent 2's incentive to resist agent 1's pressure. When the slopes are equal, the opposing forces balance each other. (If there is no point on the curve where the slopes are equal, an end point of the curve is an equilibrium point.)

Because the equilibrium point is the point of intersection with the highest intersecting power line, it represents the division that maximizes the sum of power-weighted utilities. This shows that weighted utilitarianism is also a noncooperative solution to a bargaining problem embedded in a power structure. The foregoing line of argument can be given greater mathematical precision. Also, it can be extended to general n-person bargaining problems embedded in a power structure; see Weirich (1990) for some preliminaries.

7

Application to Trustee Decisions

This chapter does not introduce a new dimension of utility analysis but applies the principles of utility analysis formulated in the preceding chapters to some problems that arise in the course of making decisions for others. Solving these problems demonstrates those principles' power, which arises from their combination. Expected utility analysis by itself is not a sufficient means of making decisions for others. Such decisions also require analysis with respect to goals and people. They require multidimensional utility analysis. Success with them vindicates the contextualist method grounding multidimensional utility analysis. It shows that contextualist decision theory is more versatile as well as more explanatory than operationist rivals.

7.1. DECISIONS FOR OTHERS

I do not treat all decisions made for others. To introduce my topic, I classify decisions made for others and identify the category I treat. I call those for whom decisions are made *clients* and distinguish decisions according to the decision maker's objective.

Parents make decisions for their children. They arrange bedtimes, menus, schools, and activities. In making these decisions they often override the beliefs and goals of their children. For example, a parent may send her child to a school that can cultivate his musical talents although the child's main goal is to attend the school his friends attend. Or a parent may insist on her child's taking an afternoon nap, although he does not believe he needs it. In general, the parent's objective is to decide in a way that promotes her child's interest, not in a way that accords with his beliefs and goals. Although attaining personal goals contributes to a child's welfare, his welfare is not constituted solely by their attainment. I call decisions for others in which the objective is the client's interest *paternalistic* decisions.

216

Physicians make decisions for their patients. They choose courses of treatment for them. In making these decisions, they often override the beliefs of their patients. For example, a physician may decide that surgery is the best means of treating a patient's cancer, although she knows her patient believes treatment by drugs would be better. The physician often has expert knowledge that warrants overriding the patient's beliefs. However, it is generally unwarranted for a physician to override a patient's goals. If the patient wants to postpone surgery because he wants to be able to attend his daughter's wedding, coming up in a month, then typically the physician should respect that goal. She should not substitute her own judgment, say, that a rapid return to health is more important than attending the wedding. She should treat the patient in a way that conforms to the patient's goals, say, by drugs until after the wedding and then by surgery. In general, the physician's chief objective should be to decide as her client would if her client had her expert information. I call decisions for others with this objective *trustee* decisions because the decision maker, the trustee, is entrusted with the task of using her expert information to serve the client's goals. Although promotion of the client's goals typically advances the client's welfare, in some cases it does not. In contrast with paternalistic decisions, trustee decisions nonetheless promote the client's goals.

Judges make decisions on behalf of legislators. They apply laws to cases that legislators have not envisaged. According to one school of thought, their objective is to decide as the legislators would if presented with the same case. They should not substitute either their own beliefs or their own goals for the legislators'. Even if a judge believes that laws of strict liability are not socially beneficial, she is not free to act on this belief in a lawsuit resting on such a law. Also, if a judge is averse to using the legal system to protect individuals from their own negligence, she is not free to act on her goals in cases where drivers break the seat belt law. When the objective of a decision for others is to decide as the client would, I call the decision a *proxy* decision because the decision maker serves as a substitute for the client.

The three categories of decisions for others are not exhaustive. They omit the anomalous decisions in which the objective is to choose according to the client's beliefs and the decision maker's goals. They also omit the commonplace decisions in which the decision maker's objective regarding the client is too imprecise for assignment to one of the three categories. Moreover, the three categories are not exclusive. Where a client's welfare and goals coincide, a decision maker may have both paternalistic and trustee objectives. However, neither exhaustiveness nor exclusiveness is necessary. A map that gives the lie of the land, if not every feature and boundary, suffices for locating the decisions I treat. According to my scheme of classification, I examine trustee decisions. They are idealized decisions in which the decision maker has a precise objective: to decide as her client would if he had her expert information.

To simplify my investigation of trustee decisions, I put aside many issues. First, conflicts of interest often arise in trustee decisions. The trustee may have objectives of her own that conflict with the objective of serving her client's interest. I put aside this issue by assuming that in the cases treated the trustee's sole objective is to serve her client's interest. This assumption also obviates a discussion of "the agency problem," the problem of creating incentives for trustees to decide in the interests of their clients.[1]

Second, questions often arise about a trustee's authority to decide on behalf of a client. Many, for example, advocate changes in health care practices to place more decision making responsibility in the hands of patients. Also, many accuse the government of unwarranted paternalism when it makes safety decisions on behalf of the public. I put aside these issues by assuming that in the cases treated the trustee is authorized to decide on behalf of the client. I assume that there are sufficient reasons for putting the decision in the trustee's hands rather than, say, sharing decision-making responsibility between the trustee and the client. For instance, it might be easier for the client to communicate his goals to the trustee than for the trustee to communicate her expert information to the client, or the client may simply lack the time required to shoulder responsibility for the decision. Moreover, I assume that in the cases treated the trustee has sufficient reasons for making a decision that serves the client's goals. In these ideal cases the client's goals are fully rational so that the objective of serving them does not conflict with the objective of serving his interests. Or, if his goals nonetheless conflict with his interests, there are good reasons for limiting the decision maker's authority to serving the client's goals, reasons related to the client's autonomy and the like. In the cases

---

[1] For a discussion of the agency problem, see Pratt and Zeckhauser (1985), Buchanan and Brock (1989), and Bowie and Freeman (1992). They consider ways of designing institutions of decision making in the professions to mitigate the effects of conflicts between the interests of professionals and their clients.

I treat, as a result, the trustee should decide as the client would if informed, and she knows this.[2]

Third, in real-life cases the trustee may lack resources needed to achieve her objective. For instance, she may lack information about the client's goals so that she does not know how to serve them. To eliminate such complications, I idealize. I suppose that the trustee is certain of the client's utility function and has unlimited cognitive abilities so that she is a fully rational ideal agent perfectly able to decide in order to serve the client's goals.

Fourth, the trustee's objective is to decide as her client would if he had her expert information. This hypothetical condition needs a sympathetic reading. I suppose that the client is a fully rational ideal agent so that he would decide in a rational way. I also assume that the imagined change in the client's information would not affect his goals so that he would decide in a way that serves his current goals. To make this more explicit, I say that the trustee's objective is to decide as her client would, given his goals and her expert information.[3] Furthermore, to simplify, I assume that if the client were informed, only one decision would maximize expected utility for him; there would not be several optimal decisions. If there were several, any

---

[2] A trustee typically does not have the objective of deciding as it would be rational for her client to decide, given his goals, if he had her expert information. A trustee typically has the objective of deciding more cautiously than her client would. For example, a banker typically endeavors to make investment decisions for clients that are more cautious than the decisions that her clients would make for themselves. Although caution exceeding the client's caution is typically a trustee's objective, it derives from factors put aside. One of these factors is a certain conflict of interest, put aside earlier along with other conflicts of interest. A trustee wants to be extra cautious to promote her own interests in the face of clients who typically blame her more for losses than they praise her for gains. Other factors put aside are the imprudence of clients' attitudes toward risk and the propriety of trustees' correcting for clients' imprudence.

[3] Certain problem cases, drawn to my attention by my colleague Robert Johnson, remain. Suppose a trustee is deciding about her client's taking steps to acquire her expert information. The appropriate objective is not the decision her client would make if he already had her expert information, not according to a natural understanding of this hypothetical decision. Also, because holding fixed the client's goals while adding the trustee's expert information may make the client's educational goals irrational, the appropriate objective is not the decision the client would make, given his goals and the trustee's expert information. To state the trustee's objective with precision, one should state it using a counterfactual conditional interpreted in terms of nearest antecedent-worlds and a specific, stipulated nearness relation between possible worlds. However such a statement of the objective would be a lengthy undertaking not crucial for my main points.

one would be an appropriate trustee decision, but I put aside ties for simplicity.

Fifth, I assume that the trustee is an expert in the sense that her information includes all the relevant information that the client possesses. Hence, if the client were to acquire her expert information, his relevant information would be the same as the relevant information she has. Furthermore, I assume that this information is sufficient to settle rational degrees of belief for states of nature relevant to the decision problem (see Section 6.4.1). I make these assumptions so that if the client were informed, his probability assignment would be the same as the trustee's in relevant respects. Thus to decide as the client would if informed, the trustee may rely on her own probability assignment.

Although my idealizations are strong, studying trustee decisions that meet them is useful. Principles governing the ideal cases serve as guides for real-life trustee decisions in the way that utilitarian principles of morality serve as guides for resolutions of real-life moral problems despite the frequent absence of the quantitative interpersonal utilities presumed by utilitarianism.

7.2. A PROBLEM CREATED BY RISK

Because trustee decisions are usually made in the face of uncertainty, a trustee needs expected utility analysis to evaluate options. Because a trustee's objective is to use her expert information to serve her client's goals, the obvious way of applying expected utility analysis is to compute the expected utilities of options using the trustee's probability assignment and the client's utility assignment. It seems that the trustee, proceeding in this way, will make the decision the client would make if he were informed, that is, if he had the trustee's probability assignment. However a problem arises.[4]

Suppose that a physician is discussing possible treatments with a patient suffering from back pain. The physician thinks that there is some chance of spontaneous remission of the pain. Nonetheless, the patient is inclined toward an operation. They use the decision tree in Figure 7.1. The patient assigns utilities to the outcomes at the terminal nodes, and the physician assigns probabilities to the states following the chance node. The expected utility of having the operation is .8, and

---

[4] This section draws on Weirich (1988b).

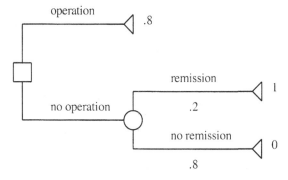

Figure 7.1. Back Pain

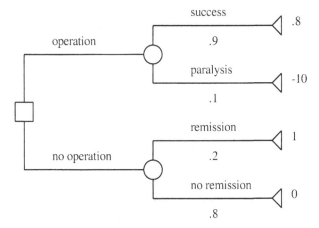

Figure 7.2. The Expanded Decision Tree

the expected utility of not having the operation is $(.2 \times 1) + (.8 \times 0) =$ .2. So having the operation maximizes expected utility with respect to this decision tree.

However, the patient's utility assignment for the operation may be uninformed. He may be ignorant of the risk of complications. Let us suppose that (a) .8 represents the utility for him of a successful operation, (b) the physician knows that there is a probability of .1 that the operation will result in paralysis, and (c) the patient's utility assignment for paralysis is −10. Expanding the decision analysis to take account of this yields the tree in Figure 7.2. The expected utility of having the operation is now $(.9 \times .8) + (.1 \times -10) = -.28$, whereas the expected utility of

221

not having the operation remains .2. Consequently, not having the operation maximizes expected utility with respect to this decision tree.

Changes in the expected utility of an option from one decision tree to another create a problem endemic to trustee decisions. Such changes do not arise in cases where a rational ideal agent makes a decision for himself. In individual decisions all trees yield the same value for the expected utility of an option because the utilities at the terminal nodes of a tree equal the probability-weighted averages that replace them in expansions of the tree. In trustee decisions the possibility that an option's expected utility changes with a decision tree's expansion arises because the trustee, rather than the client, is the source of the probabilities for the states added. The expansion simulates a change in the client's information concerning those states. It is as if the trustee's information about those states becomes the client's. Of course, expected utilities may change when information changes.

The example shows that when applying expected utility analysis to a trustee decision, the decision tree must be carefully constructed. It must be elaborated until the client utilities at terminal nodes are *stable*, that is, until they would not change if the client were to acquire the trustee's expert information. Otherwise the tree may not produce the decision the client would make if informed.

Unfortunately, expanding the decision tree is not always enough to eliminate unstable client utilities. To show this, I begin by distinguishing two factors that cause the client utilities at terminal nodes to change as the client's information changes. The first factor is the client's uncertainty about events beyond the terminal node. This uncertainty causes the instability of utilities exhibited by the example. It can be accommodated by adding branches that represent the events about which the client is uncertain. After those branches are added, uncertainty about the events no longer influences the utilities at terminal nodes. When a decision tree is expanded to accommodate all relevant uncertainties, I call it a *maximally expanded* tree. In the example, if the second tree depicts all the relevant possibilities, it is a maximally expanded tree.

The second factor that causes unstable client utilities at terminal nodes is instability in the outcome of an option at a terminal node. The client utility at a terminal node has the form $U(o$ given $(s_i$ if $o))$, where o is an option and s_i is a state comprising all the states leading to the terminal node. Its value depends on the outcome of o given $(s_i$ if $o)$. Most components of the outcome are independent of the client's information. But the (subjective) risk involved in the option is not. The risk depends, for example, on the probabilities and utilities of the possible outcomes of the option, as Section 4.1 explains. Because the probabilities may change as the client's information changes, the risk may change as well. Of course, if the outcome at a terminal node changes as the client's information changes, then the utility at the terminal node may also change.

The problem may be stated in terms of possible worlds (trimmed of irrelevant detail). The world at the end of a maximally expanded branch for an option includes the option itself and whatever risk that option involves. The risk the option involves depends on the client's probability and utility assignments, among other things. Thus the client's utility for the world reflects the client's attitude to the risk the option involves, as that risk is assessed with respect to his own probability assignment. If the client were to acquire the trustee's information, he may have a new probability assignment. The risk involved in the option may change. So the world at the terminal node may be different. It may be a world in which his undertaking the option has a different risk for him. For him, the utility of this new world may differ from the utility of the original world. Thus although the utility of worlds is stable, the world at the terminal node of a decision tree depends on the client's risk in undertaking the option, and so may change as he acquires the trustee's information. Therefore the utilities at the terminal nodes of a maximally expanded tree may not be stable. Because of risk, even a maximally expanded decision tree may not produce the decision the client would reach if informed.

To illustrate the problem, suppose that in the example the patient is aware of the expanded decision tree in Figure 7.2 and thinks that the probability the physician assigns to a successful operation is based on a large sample, whereas the physician knows that it is based on a small sample. Then if the patient were to acquire the physician's information, the operation would seem riskier to him, although the relevant probabilities do not change. As a result, the utilities at the terminal nodes for its branch would decrease. This may happen even if the tree is maximally expanded. Although expanding the decision tree can handle the instability of utilities that arises from uncertainty about events, it cannot handle the instability of utilities that arises from outcomes that vary with information. Because of components of outcomes such as risk, a maximally expanded tree might not produce the decision the client would make if informed.

There are various ways of solving the problem with risk. The simplest way is to assume that the trustee knows directly the utility assignment the client would make if he were informed. Then the trustee can use this hypothetical utility assignment to obtain suitable utilities for the terminal nodes of any decision tree she uses. For the sake of practicality, however, it is best to solve the problem while assuming only direct knowledge of client utilities that are easily accessible to the trustee – that is, client utilities that are easy for the client to communicate to the trustee. I try to solve the problem using hypothetical client utilities selectively. I use ones easy to obtain directly to compute ones more difficult to obtain directly.[5]

My approach supplements expected utility analysis with a method of computing stable client utilities for terminal nodes from the client's utility assignment for outcomes of options. As an additional dimension of utility analysis, the computation uses goals, more precisely, basic intrinsic desires and aversions – that is, the fundamental desires and aversions that cause all other desires and aversions. It applies partial utility analysis, which Section 5.3 introduces and Section 5.4 illustrates, and so involves utility functions evaluating a proposition with respect to partial sets of basic intrinsic attitudes. My applications use utility functions with respect to basic intrinsic aversions to risks and with respect to all other basic intrinsic attitudes taken together. Thus they assume that the client has basic intrinsic aversions to risks. This is not a strong assumption because aversions to risks are causally fundamental in many cases.

My analysis of the utility of an option o begins with a maximally expanded decision tree. It seeks an informed utility assignment for the option's outcome at a terminal node generated by a series of states. The outcome may be expressed by the proposition that o given (s_i if o), where s_i represents the combination of states leading to the terminal node. To abbreviate, I write o_i. The analysis then splits the evalua-

---

[5] To put aside unnecessary complications, I do not distinguish between hypothetical and conditional utility assignments – that is, assignments made if a condition holds and assignments made given the assumption of a condition.

Also, notice that because in trustee decisions beliefs spring from one source, the trustee, and desires spring from another source, the client, the relevant probabilities and utilities cannot be inferred from the trustee's preferences using the techniques exhibited by proofs of familiar representation theorems – not unless the trustee is assumed to take on the client's preferences, which requires giving the trustee information about the client's preferences as well as the objective of honoring the client's preferences.

tion of o_i into two parts: an evaluation with respect to risk, $U_r(o_i)$, and an evaluation with respect to factors besides risk, $U_{r\sim}(o_i)$. The second utility, $U_{r\sim}(o_i)$, is informed because the specification of the outcome at a terminal node of a maximally expanded tree explicitly resolves every relevant uncertainty. In contrast, the first utility, $U_r(o_i)$, is not informed because it depends on the client's probability assignment and assessment of the risk involved in o. Hence the analysis substitutes an informed partial utility obtained from (1) the risk assessed using the trustee's information and probability assignment, and (2) the client's aversion to that risk. The informed utility of the outcome is then the sum of two informed partial utilities.

More precisely, let U' be the client's utility assignment conditional on the trustee's assessment of the risk involved in o, a risk independent of the state s_i. If risk is the only information-sensitive consequence of o neglected by the maximally expanded tree, $U'(o_i)$ is the informed utility sought. It takes account of the trustee's information about crucial points, appealing to the trustee's probability assignment and other relevant information when assessing the risk involved in o. It is a stable client utility for an expected utility analysis using the maximally expanded tree. Also, because $U'(o_i)$ involves a simple condition concerning risk, it can be calculated in a practical way. By partial utility analysis, $U'(o_i) = U'_r(o_i) + U'_{r\sim}(o_i)$. The summands can be obtained from modest direct information about the client's utility assignment. Because o_i is at the terminal node of a maximally expanded tree, $U'_{r\sim}(o_i) = U_{r\sim}(o_i)$. Because both the condition on which U' rests and the basic intrinsic attitude on which the partial utility U'_r rests concern risk exclusively, $U'_r(o_i)$ is just the strength of the client's aversion to the risk involved in o as the trustee assesses that risk, something the trustee can calculate, given knowledge of his client's attitudes toward risks.

As the foregoing shows, partial utility analysis used in conjunction with expected utility analysis yields a satisfactory trustee decision. Partial utility analysis applied to an outcome's utility solves a problem that expected utility analysis alone cannot handle in a practical way because it cannot separate an outcome's utility into evaluations with respect to partial sets of basic intrinsic attitudes.

To illustrate the hybrid form of utility analysis, imagine that a physician must choose between two courses of treatment for a patient's disease. One course of treatment is old and well tested. The other is new and relatively untested. Both offer the same probability of a cure. Putting aside risk, the decision problem generates the tree in

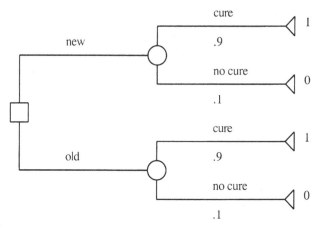

Figure 7.3. A Trustee's Analysis of Risk

Figure 7.3, which I suppose is maximally expanded. Because the new treatment is untested, the risk it involves is greater than the risk the old treatment involves. A risk is bigger the less evidence there is to support the probabilities it involves; information reduces risk. Because $U'_r(o_i)$ responds to the trustee's assessment of the risk involved in o, it is greater, for each state i indexes, if o represents the old treatment than if o represents the new treatment. Because a cure under each treatment receives the same evaluation putting aside risk, and similarly for a failure to achieve a cure, partial utility analysis yields informed utilities at terminal nodes that are higher for the old treatment than their counterparts for the new treatment. Consequently, applying the hybrid analysis, the old treatment has higher expected utility than the new treatment. It is the rational trustee decision.

This example presumes that the patient is averse to risk, which is typical in such cases. People are sometimes attracted to risk, however. Kahneman and Tversky (1979: 278) observe that people are risk-seeking when confronted with probable losses. They are attracted to risks that offer a chance to escape from doom. Imagine, then, a new version of the example that reverses the probabilities for *cure* and *no cure*. Instead of both treatments offering a good chance of a cure, both offer little chance of a cure. In this case patients typically are risk seekers rather than risk averters. A typical patient will find the riskier treatment more attractive, other things being equal. Because the new treatment is less well investigated, its improbability of curing the disease is less well established. It therefore offers more hope. A typical

226

patient wants to take on its greater risk in the hope of ending the disease. The options are still tied with respect to factors besides risk, and the new treatment is still more risky than the old treatment, but now its greater risk is an advantage that tips the balance in its favor. Its greater risk yields greater informed utilities for outcomes. It maximizes expected utility for the informed patient, so the hybrid method of utility analysis favors it. It is the rational trustee decision given the probability reversal.

This conclusion presumes the chapter's idealization to cases in which a trustee should serve her client's objectives without attempting to compensate for mistakes. Under this idealization, a rational trustee decision responds to a client's attraction to risk even if it is mistaken. However, the conclusion may stand without that idealization. Attractions to risk need not be mistaken. Standards of rationality tolerate variation in attitude toward risk. This is plainly the case for standards of instrumental rationality. It is plausible also for substantive standards of rationality. The variation in attitude that Kahneman and Tversky observe is not at odds with any well established standard of rationality. Although substantive standards of rationality may constrain attitudes toward risk, they do not prohibit all attraction to risk. One virtue of applying this section's hybrid utility analysis to trustee decisions is that it tolerates diversity of attitudes toward risk. It does not build into utility analysis any particular attitude toward risk.

When using partial utility analysis to compute stable client utilities for trustee decisions, I have assumed that $U'_r(o_i)$ is obtained on a case-by-case basis. I have assumed that the trustee has direct knowledge of the risk involved in o_i, given her probability assignment and other relevant information, and has direct knowledge of the client's basic intrinsic aversion toward that particular risk. I have not advanced a general formula for computing $U'_r(o_i)$ from more accessible information. The assumptions allow for variability in the way that the client assigns utilities to risks. However, a formula for calculating $U'_r(o_i)$, even if it applies only to a restricted class of cases, may further reduce the difficulty of acquiring direct knowledge sufficient for calculating informed client utilities. One approach is to calculate $U'_r(o_i)$ from the size of the risk involved in o according to the trustee's assessment and the strength of the client's basic intrinsic aversion to risks of that size. This calculation requires a quantitative measure of risk. Several have been proposed although each is controversial; see Libby and Fishburn (1977). The calculation also assumes that the strengths of the client's basic intrinsic aversions toward risks are sensitive only to the sizes of the risks. This assumption limits its application but nonetheless allows for an interesting range of cases. Elaboration of this method of calculating $U'_r(o_i)$ is left for future research.

The overall method of handling risk may also be generalized for cases where trustees have ample information about their clients. Because partial utility analysis is completely general, the method does not require the maximally expanded trees in terms of which it was introduced. A trustee may use partial utility analysis to divide the utility of an outcome at a terminal node of any decision tree into its utility with respect to risk and its utility with respect to other factors. For an arbitrary proposition p, partial utility analysis yields the formula $U'(p) = U'_r(p) + U'_{r\sim}(p)$. So a trustee may use this formula to obtain a client's utility for an outcome at a terminal node of an abbreviated decision tree, provided she knows the requisite partial utilities. Any decision tree for which the trustee has appropriate information makes a suitable framework for an application of the hybrid method of utility analysis.

7.3. TRUSTEE DECISIONS WITH GROUPS

Next, I address trustee decisions in which either the trustee or the client is a group of individuals rather than a single individual. My enriched decision theory can handle the additional complexities. As usual, I make some idealizations to apply it. Results given those idealizations guide deliberations in more realistic cases.

7.3.1. Clients That Are Groups

Suppose that a government official is making a decision for the public, perhaps a decision about the regulation of some toxic substance. Also, suppose that the official's objective is to make the decision the public would make for itself if it were informed. Then the official is making a trustee decision in which the client is a group of individuals, the public. Multidimensional utility analysis helps trustees make decisions on behalf of groups in such cases.

When a trustee's client is a group, deciding as the group would if informed is an appropriate objective. I admit, of course, that this objective may be overridden by other appropriate objectives. For instance, the group if informed might decide in a way that is unfair to some of its members. Then the trustee may have an obligation to foil the group. In the following, to put aside cases where a trustee's objectives conflict, I suppose that the group would reach a fair decision if informed. In particular, if differences in power among the group's members would influence its decision, that influence would be fair. Considering this, I continue to assume that the trustee's sole objective is and should be to decide as the client would if informed.

Note that even in cases where the assumption is not met and a trustee has other objectives, she may want to know the decision the group would make if informed. She may not be authorized to take account of her other objectives, or to take account of them she may need the group's informed decision as a bench mark. Thus my method of making a trustee decision for a group, calculating the group's informed decision, is relevant in cases besides those treated here.

When I speak of the decision a group would make if informed, I mean more fully the decision it would reach through informed negotiations given some idealizations. First, the group's members are fully rational. Second, negotiations are unconstrained, that is, communication is effortless and instantaneous, and without cost the group's members can make binding contracts providing side payments for concessions. Third, the group's members have quantitative interpersonally comparable utility assignments and stand in quantitative relations of social power to each other. Under these idealizations, the group reaches a decision that maximizes group utility as defined in Chapter 6. That is, it maximizes the sum of power-weighted interpersonal utilities.

The weighted utilitarian outcome assumes that all the group's members have the same probability assignment. This assumption is met under the hypothesis that the group is informed, that is, each member has the trustee's relevant information. I assume that the trustee is an expert in the sense that her information contains the relevant information of the client. When the client is a group, this amounts to the assumption that the trustee's information includes the relevant information of each member of the group. As a result, if the group were informed, that is, if every member had the trustee's relevant information, every member's relevant information would be the relevant information she has. Moreover, I assume that the trustee's relevant information is sufficient to settle a rational degree of belief for each relevant state. Hence, if every member of the group had her relevant information, all would have her probability assignment. It follows from

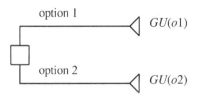

Figure 7.4. Group Utilities

these assumptions that the group's members would have the same probability assignment if informed.[6]

Having stated its assumptions, I now apply my method of trustee decision making to cases where the client is a group and its utility assignment is obtained by group utility analysis. The main tactic, carried over from Section 7.2, is to maximize expected utility using a decision tree in which the probabilities of states are trustee probabilities and the utilities at terminal nodes are informed group utilities. To illustrate, consider the tree in Figure 7.4, where chance nodes are suppressed for simplicity. The utilities at terminal nodes are informed group utilities of the options. These may be computed by group utility analysis from informed utilities of the options for the group's members. Suppose that the group has two members and the first is twice as powerful as the second. Let $U_1(o_1)$ be the informed utility of the first option for the first member of the group, and so on. Then the group utilities are as follows: $GU(o_1) = 2U_1(o_1) + U_2(o_1)$, and $GU(o_2) = 2U_1(o_2) + U_2(o_2)$.

Because the informed probability assignment for each member of the group is the trustee's probability assignment, one may use the trustee's probability assignment for states when applying expected utility analysis to the informed group utility at the terminal node of a tree. To illustrate, I expand Figure 7.4's tree at the top terminal node in order to apply expected utility analysis to the group utility of the first option. The result is in Figure 7.5. As before, the group utilities at terminal nodes are informed group utilities. According to expected

---

[6] Chapter 6 advances weighted utilitarianism for all group decisions in ideal circumstances, but a supporting argument in Section 6.5 treats only bargaining problems embedded in a power structure. To obtain a trustee decision problem that builds on that argument's foundation, imagine that the trustee is an arbitrator enlisted to resolve a group's bargaining problem. To simplify the argument for the weighted utilitarian outcome in these circumstances, also imagine that the group's bargaining problem is embedded in a static power structure, one unaffected by potential outcomes.

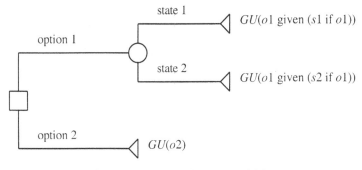

state 1

$GU(o1$ given $(s1$ if $o1))$

option 1

state 2

$GU(o1$ given $(s2$ if $o1))$

option 2

$GU(o2)$

Figure 7.5. Expected Group Utilities

utility analysis, $GU(o_1) = P(s_1$ given $o_1)GU(o_1$ given $(s_1$ if $o_1)) + P(s_2$ given $o_1)GU(o_1$ given $(s_2$ if $o_1))$. Here the probabilities of the states are the trustee's; they are the informed probability assignments of every group member.

Sometimes a trustee deciding for a group will want to use group utility analysis to compute the informed group utility of an option by first supposing that the members become informed, and then calculating the power-weighted average of the informed utilities of the option, as the group utility of option 1 was computed for Figure 7.4. This method is convenient for small groups. Other times the trustee will want to use expected utility analysis to compute the informed group utility of an option from her probability assignment for states and informed group utilities for option-state pairs, as the group utility of option 1 was computed for Figure 7.5. This method is convenient for large groups. Section A.6 shows that the two procedures are equivalent; both partition the reasons for an informed group utility assignment. Group and expected utility analyses may be used separately or in combination for trustee decisions.

Also, to obtain informed group utilities at terminal nodes, partial utility analysis is often useful. It can introduce partial utilities of outcomes to take account of aversion to risk, as in Section 7.2. Suppose, for instance, that every member of the group has a basic intrinsic aversion to risk. Then the group has a basic intrinsic aversion to risk. Partial utility analysis applies as it does for individuals. It splits the group utility for an outcome into a component due to risk and a component due to other factors. In a maximally expanded decision tree the informed group utility for the outcome o_i generated by an option o and series of states s_i is $GU'(o_i)$, where GU' is the group's utility assignment

231

given the trustee's assessment of the risk involved in the option. By partial utility analysis, $GU'(o_i) = GU'_r(o_i) + GU'_{r\sim}(o_i)$, where GU'_r is group utility with respect to risk and $GU'_{r\sim}$ is group utility with respect to factors besides risk.

Also, when making decisions for a group, a trustee may use intrinsic utility analysis in conjunction with group utility analysis because both derive from the same source, the principle of pros and cons. In fact, intrinsic utility analysis may be used simultaneously with both expected and group utility analyses; the various forms of analysis are just different ways of partitioning the same reasons for informed group utility assignments (see Section A.6). Consequently, a trustee has great latitude in choosing a procedure for calculating the informed group utilities of options. She may apply expected, group, and intrinsic utility analyses in any order that is convenient.

7.3.2. Trustees That Are Groups

Next, consider trustee decisions in which the trustee is a group of people. Some examples are a government agency's decisions for the public, a corporation's board of directors' decisions for the shareholders, and a faculty committee's decisions for the university community. In these examples the client is a group but so is the decision maker. Such cases pose a new complication for the decision procedure advanced. What is the trustee's probability assignment? Evidently, it should be defined in terms of the probability assignments of the people composing the trustee body. But the appropriate definition depends on the circumstances.

My methods of making trustee decisions are proposed under idealizations, roughly to the effect that conditions are perfect for meeting the trustee's objective to decide as the client would if informed. In cases where the trustee is a group, I take the idealizations to include the following. First, the trustees are fully rational ideal agents. Second, the trustees can communicate effortlessly and instantaneously, and the information available to them, taken together, is sufficient for settling a probability assignment for the states relevant to their decision. Third, the trustees' sole objective is making an informed decision for the client. So they are willing to share information and use the probability assignment that results for their decision. They do not need compensation for sharing information. Under these idealizations it is appropriate to let the trustee probability assignment be the trustees'

232

common probability assignment after pooling information and applying inductive logic. In other words, the trustee body's probability assignment is the group probability assignment defined for the trustees by Section 6.4.1.

The assumptions about a collective trustee dovetail with the previous section's assumptions about a collective client. For a collective client, it assumes the opportunity for cost-free, instantaneous communication and an incentive to share information. These assumptions facilitate the definition of a group utility assignment because they provide for a common underlying probability assignment. For a collective trustee, this section makes similar assumptions to obtain a common probability assignment to serve as the trustee's assignment. As a result, the trustee's assignment is the assignment a collective client would adopt if informed; it is the assignment that all the group's members would adopt if informed. Hence it underlies a client's informed utility assignment, whether the client is an individual or a group.

Once the appropriate trustee probability assignment has been identified, my methods of trustee decision making apply straightforwardly to cases in which trustees are groups. They recommend a decision that maximizes expected utility with respect to a decision tree in which states are assigned trustee probabilities and possible outcomes are assigned informed client utilities. The full resources of multidimensional utility analysis are available for dissecting utilities in any way that is convenient.

7.3.3. An Illustration

To illustrate the application of my enriched form of decision theory to trustee decisions involving groups, I review a regulatory decision by the Occupational Safety and Health Administration (OSHA). To apply my theory, I assume the idealizations previously specified, for instance, the idealization that everyone involved is a fully rational ideal agent. Because the real world falls short of the idealizations, what follows is merely an illustration of the theory and not a practical application of the theory to a regulatory decision.[7]

---

[7] For a more realistic treatment of regulatory decisions, see Fischhoff et al. (1981), Mayo and Hollander (1991), Cranor (1993), and Hahn (1996).

To start, notice that a regulatory decision is made by the government for the public on a matter about which the public lacks technical expertise. The government's responsibility is to make an informed decision serving the public's goals. Regulatory decisions are ones in which the public's goals are to be respected, but the public's beliefs about technical matters are to be replaced by experts' beliefs. This distinguishes regulatory decisions from electoral decisions, where both the public's beliefs and goals are to be respected. The sample regulatory decision assumes that the government's objective is the decision it would be rational for the public to reach via unconstrained negotiations if it were informed.

Regulatory decisions are made in a complex way. Congress has the responsibility of using available information to make decisions on behalf of the public. However, it delegates some of its responsibilities to the regulatory agencies. Specifically, it instructs the agencies to survey information on hazardous technologies and form an expert opinion about the risks involved. It also instructs the agencies to use this expert opinion and the public's goals, as articulated by Congress in regulatory directives, to decide on standards for the technologies. In particular, OSHA has the job of using technological information to impose standards that reduce workplace hazards. Because the primary regulatory body, Congress, delegates its epistemic and decision-making responsibilities, regulatory decision making by administrative agencies is bound by delegatory institutions. However, I put aside institutional obligations and simply assume that the regulatory agencies should aim for decisions the public would make if informed.

Because the public and the regulatory agencies are groups, regulatory decision making must address the amalgamation of the opinions and desires of individuals to form the opinions and desires of groups. I handle the amalgamation of opinions through the method of Section 7.3.2, namely, pooling information and applying inductive logic. I handle the amalgamation of desires through group utility analysis as in Section 7.3.1. To facilitate group utility analysis, I assume that a regulatory decision by the public may be treated as a bargaining problem embedded in a static power structure so that weighted utilitarianism applies noncontroversially (see Section 6.5).

My example is OSHA's decision in 1978 to reduce exposure to benzene. Benzene is a carcinogen that causes leukemia. It is produced when petroleum is refined. It is present in gasoline and is used as a solvent, for example, in manufacturing automobile tires. According to available information, there is no safe level of exposure to benzene. Any level of exposure may cause cancer. Exposure to benzene in the petroleum industry can be reduced by respirators at petroleum refineries and by gloves when using benzene as a solvent. Hence OSHA imposed new regulations reducing permissible levels of workers' exposure to benzene.

The new regulatory standard was challenged by the American Petroleum Institute and the issue went to the Supreme Court. Debate in the Supreme Court focused on two regulatory principles appearing in the Occupational Safety and Health Act of 1970. The principles concern conditions for OSHA's imposition of safety and health standards. Both principles presume that OSHA imposes standards regulating the most serious hazards first so that the resources available for regulating hazards are used optimally.

The first regulatory principle was drawn from the definition of the occupational safety and health standards that the act orders. Here is the definition provided by the *United States Code* (1995: 16:152).

The term "occupational safety and health standard" means a standard which requires conditions, or the adoption or use of one or more practices, means, methods, operations, or processes, reasonably necessary or appropriate to provide safe or healthful employment and places of employment. (29 USC sec. 652(8))

As interpreted in the Supreme Court dispute, this passage, taken with the rest of the act, orders workplace standards that are economically feasible and will prevent some cases of serious injury or impairment of health in the foreseeable future.

The second principle was drawn from the act's discussion of standards for toxic materials. The relevant passage from the *United States Code* (1995: 16:154) follows.

The Secretary, in promulgating standards dealing with toxic materials or harmful physical agents under this subsection, shall set the standard which most adequately assures, to the extent feasible, on the basis of the best available evidence, that no employee will suffer material impairment of health or functional capacity even if such employee has regular exposure to the hazard dealt with by such standard for the period of his working life. (29 USC sec. 655(b)(5))

As interpreted in the Supreme Court dispute, this passage orders workplace standards that are economically feasible and have a significant chance of preventing some cases of serious injury or impairment of health in the foreseeable future.

How do these two principles apply to standards regulating workplace exposure to carcinogens? The arguments presented to the Supreme Court targeted outcomes concerning cancer prevention and the cost for the petroleum industry of implementing safety and health standards. They appealed to the magnitude and likelihood of these outcomes to defend or attack OSHA's new standard for benzene. They granted the principles' common presupposition that the public strongly prefers cancer prevention for workers to protection of corporate profits. In applying the principles, I similarly assume that the public has this preference and that the preference is informed and stable. I use the public's attitudes toward cancer and the costs of preventive measures, together with relevant expert information, to compare these two principles for making choices between standards to promote cancer prevention and strict limits on government regulations.

The main difference between the two principles, for carcinogens such as benzene, is that the first orders standards that will prevent some cases of cancer, whereas the second orders standards that have a significant chance of preventing some cases of cancer. For convenience, I call the first principle "the principle of guaranteed benefits" because it requires showing that a standard will bring some health benefits. I call the second principle "the principle of likely benefits" because it requires showing that a standard has a significant chance of bringing some health benefits.

Each party in the Supreme Court debate appealed to one principle and attempted to finesse the other. OSHA invoked the principle of likely benefits. It argued that the standard proposed for benzene has a significant chance of preventing some cases of cancer, because it decreases exposure to benzene, a carcinogen for which no safe level of exposure is known. The petroleum industry invoked the principle of guaranteed benefits. It argued that the evidence that the standard would prevent some cases of cancer is inconclusive, because, as far as anyone knows, the present levels of exposure to benzene are already safe. The Supreme Court did not contest the applications of the principles but gave precedence to the principle of guaranteed benefits, and therefore rejected OSHA's new standard.

I do not use this chapter's methods of trustee decision making to argue for any particular regulations concerning benzene. My objective is more limited. I use these methods to argue that neither the principle of guaranteed benefits nor the principle of likely benefits is a suitable regulatory principle for the case. The problem with the principle of guaranteed benefits is easy to see. It requires that the imposition of a standard for benzene be supported by evidence that the standard will prevent some cases of cancer. However, typically, as with OSHA's standard for benzene, it is not clear that a standard will prevent cases of cancer. Rather the evidence only makes it likely that the standard will prevent some cases of cancer. Insisting on practical certainty that the standard will prevent some cases of cancer means waiting for present conditions to cause some people to fall victim to cancer. A regulatory principle should be able to authorize action before that happens.

The principle of likely benefits creates the opposite problem; it authorizes the imposition of regulations too readily. It authorizes the reduction of risks no matter how small and no matter how unlikely to result in injuries. It lays itself open to the charge of unrealistically seeking a risk-free workplace.

Proponents of the principle of likely benefits would, I think, argue for it in the following way. First, they would claim, as is common, that the utility of a chance for a benefit is the utility of the benefit discounted by the probability of receiving it. Then they would claim that, for the public, the utility of preventing cases of cancer for workers is very high compared with the utility of protecting corporate profits. From this they would conclude that the utility of any significant chance of preventing cases of cancer for workers is high compared with the utility of protecting corporate profits. Therefore, following my methods of trustee decision making, they would claim that a likely reduction in cases of cancer justifies the imposition of any economically feasible standard.

This argument's claim about the utility of a chance for a benefit appeals to the expected utility rule according to which, if states are independent of options, the expected utility of an option o is $\Sigma_i P(s_i) U(o$ given $s_i)$. The argument furthermore assumes that $U(o$ given $s_i)$ may be computed without counting risk (broadly construed as taking a chance) even if o is chancy. As Section 4.1 shows, however, the utility of the risk o creates plays a role in fixing the value of $U(o$ given $s_i)$. In fact, if we assume that the public has a basic intrinsic aversion to risk, its utility assignment conforms to this equation: $U(o$ given $s_i) = U_r(o$ given $s_i) + U_{r-}(o$ given $s_i)$, where U_r is partial utility taken with respect to risk and

$U_{r\sim}$ is partial utility taken with respect to other factors (see Section 7.3.1). As a result, the argument for the principle of likely benefits falters.

Suppose that o is the imposition of a certain standard for a carcinogen and that s_1 is the set of circumstances in which the standard prevents some cases of cancer. For a standard with a significant chance of preventing some cases of cancer, the probability discounted utility of the benefit is high relative to the utility of protecting corporate profits. Hence $P(s_1)U_{r\sim}(o$ given $s_1)$ is large. However, $U_r(o)$, the utility of the element of chance involved in imposing the standard, is typically negative and large for three reasons. First, the public is averse to taking chances, at least when seeking possible gains. Second, standards typically have large costs so that the dispersion of possible gains and losses is large and negatively skewed. Third, the weight of the evidence that the standard has a significant chance of preventing some cases of cancer is typically low because (a) there is no well-established dose-response curve valid in the neighborhood of the standard, (b) the positive evidence that the standard has a significant chance of preventing some cases of cancer relies on extrapolation, and (c) there is some countervailing evidence. Since $U_r(o)$ is large and negative, $U_r(o$ given $s_1)$ is also large and negative because the risk involved in a chancy option is the same regardless of the state determining the option's outcome. Thus $P(s_1)[U_r(o$ given $s_1) + U_{r\sim}(o$ given $s_1)]$ may be a small positive quantity, or even a negative quantity. As a result, the utility of the standard's chance of preventing some cases of cancer may not offset the costs of the standard.

One possible objection to this criticism of the argument for the principle of likely benefits is that aversion to taking chances counts against doing nothing to eliminate risks of cancer as well as against standards with only likely benefits. Hence, even considering aversion to taking chances, the public's preference for cancer prevention over protection of corporate profits still supports the principle of likely benefits.

My criticism, however, does not deny that aversion to taking chances affects the evaluation of doing nothing to reduce risks of cancer. It merely claims that no matter how offensive present risks of cancer are, the utility of a reduction in those risks has to be discounted in view of an aversion to the chance that the reduction in risk will not actually prevent any cases of cancer. After the utility of the reduction has been suitably discounted, the utility of the reduction may not be positive.

The foregoing shows how my analysis of trustee decision making sheds light on regulatory decisions. It shows that the principles of guaranteed benefits and likely benefits are both unsatisfactory. Both omit considerations that regulatory decisions should take into account. I hope the analysis will provide a framework for the formulation of better principles of regulation.[8]

7.4. THREE-DIMENSIONAL UTILITY ANALYSIS

This chapter shows that, given uncertainty, trustee decision making demands the power of multidimensional utility analysis. When the client is averse to risk, it needs the dimensions of basic intrinsic attitudes and possible outcomes. When the client is a group, it needs the

---

[8] This discussion of regulatory legislation does not question the propriety of society's treating worker safety through the Congress. However, some points argue for a different approach. The Congress applies representative democracy to issues. Its objective with respect to an issue is the decision the public would make for itself if informed. However, suppose the public's decision on worker safety would ignore the rights of workers because a majority of citizens seeks inexpensive products. Then a congressional resolution of the issue would be unjust toward workers. Granting that workplace safety is a matter of individual rights, it should be recognized as a constitutional right to be protected by the judiciary branch of government. The Constitution has the function of protecting individual rights against assaults from a majority.

If a constitutional right to workplace safety is recognized, then as a first approximation to a judicial principle elaborating that right I suggest a principle calling for safety standards that a reasonable, informed, representative worker would accept. A representative worker has only a moderate aversion to risk. An informed worker has all the relevant information available. And a reasonable worker balances safety against productivity and related matters such as pay and employment opportunities. Without introducing technical material that would make it unwieldy in the public arena, this principle furnishes a guideline for a just balance between worker safety and productivity, a guideline for responsible risk taking. It may be supported by a contractarian argument.

This judicial principle yields different safety standards than my principle of trustee decision making. The preferences of a representative worker are an average of the preferences of workers, but they are not a power-weighted average. Hence, other things being equal, safety standards arising from the judicial principle are more egalitarian than safety standards arising from my principle of trustee decision making.

Viscusi (1998: 64) advances a principle similar to the judicial principle. He says, "The appropriate basis for protection is to ask what level of risk workers would choose if they were informed of the risks and could make decisions rationally." He wants to remind affluent policy makers and risk analysts of the workers' viewpoint, and does not necessarily have in mind the reasons behind the judicial principle, however.

dimensions of people and possible outcomes. And when the client is a group and is averse to risk, it needs all three dimensions of utility analysis. Also, the flexibility of multidimensional utility analysis allows a trustee to apply it despite limited information. Because intrinsic, expected, and group utility analyses are fully compatible in virtue of their common derivation from the principle of pros and cons, they may be applied in any order convenient for the trustee. Multidimensional utility analysis makes trustee decisions tractable.

8

Power and Versatility

Multidimensional utility analysis is a versatile tool for decision making. It creates a decision space with room for many varieties of utility analysis. A decision maker may employ whichever analysis best suits the problem she faces.

My version of multidimensional utility analysis accommodates various traditional forms of utility analysis and ensures their consistent conjoint application. Its cornerstone is intrinsic utility analysis. My new formulation of it establishes its credentials and guides the formulation of other methods of utility analysis. It refines those methods of utility analysis to make them more accurately handle considerations such as aversion to risk.

Multidimensional utility analysis requires a decision space in which to locate and separate reasons for options. Given the idealizations about agents and their decision problems, the finest-grained reason is a chance for realization of a person's basic intrinsic attitude. Such reasons occupy the points of my decision space. When agglomerated various ways, they yield an option's utility. The forms of agglomeration furnish my methods of utility analysis. Intrinsic, expected, and group utility analyses fully survey reasons for options.

The decision space I adopt not only makes precise and unifies traditional forms of utility analysis, but also suggests new forms of utility analysis such as partial utility analysis. The new forms of analysis present new ways of organizing the reasons behind traditional utility analyses. They make multidimensional utility analysis even more versatile.

My approach to utility analysis relies on nonoperationist, contextualist methods of introducing concepts. It draws on the strength such methods give decision theory. The enriched theory they produce handles complex trustee decisions with precision and subtlety.

The decision rules and utility principles I advance rest on idealizations. Some idealizations treat agents. Others treat agents' situations. The idealizations for groups are especially powerful because they must justify treating groups as agents to which norms of individual rationality apply. Realism calls for the eventual removal of idealizations. An important project for future research is the extension of multidimensional utility analysis to nonideal agents in nonideal situations. Although discharging idealizations will force adjustments in the decision rules and utility principles, the rules and principles advanced here are useful guides to practical decision making and are significant steps toward new principles even more practically useful because less reliant on idealizations.

One may wonder about the scope of multidimensional utility analysis. I have restricted it to familiar dimensions of utility analysis: the basic dimensions of goals, outcomes, and people. Are there other dimensions to add? That is a good topic for future research. I hope my framework suggests fruitful avenues of investigation. One expansion briefly sketched includes auxiliary dimensions of time and space. Temporal and causal forms of utility analysis using these dimensions put aside certain irrelevant reasons in the past and otherwise outside an agent's control. They simplify options' evaluations. Although the outlines of these forms of analysis are clear, details await progress with theories of causation and the boundaries of events.

A related question concerns utility. Are there principles constraining it besides the ones appearing in traditional and allied forms of utility analysis? For example, are there principles regulating aversion to risk, or requiring a present concern for satisfaction of future desires, or requiring a present concern for execution of a plan adopted earlier? These questions raise deep issues. But however they are resolved, they do not affect my framework for utility analysis. Such additional constraints, if warranted, operate within the framework. They affect the utility assignments multidimensional utility analysis dissects, but they do not alter its methods.

A final question about scope addresses decision principles using utilities as input. The idealizations allow reliance on the rule to maximize utility. As those idealizations are removed, do other decision principles emerge? There may be cases where utility maximizing should give way to alternative principles such as satisficing. If so, do these substitutes force revisions of utility analysis? Again, this is a controversial matter.

But if my framework for utility analysis is well designed, its modification for nonideal cases will support my decision principles generalized for nonideal cases. Confirming the framework's endurance is an important future project.

Appendix: Consistency of Calculations of Utilities

I have introduced many ways of computing utilities. Are they consistent? For example, is an expected utility analysis of the comprehensive utility of an option, $U(o)$, consistent with its intrinsic utility analysis? I have already argued for the correctness of the various forms of utility analysis. Because correctness entails consistency, I have already argued indirectly for consistency. In particular, the consistency of the forms of utility analysis follows by their common derivation from the principle of pros and cons. Still, I would like to verify consistency independently to check my applications of the principle.

Another related issue is whether forms of utility analysis for one type of utility apply to other types of utility. For example, does an option's group utility as defined in terms of its utility for group members equal its value according to an expected utility analysis? This is an issue of compatibility rather than consistency. But if there is a standard method of calculating each type of utility, the issue may be reduced to the consistency of a standard and a hybrid calculation for a type of utility. Furthermore, given two hybrid utility analyses of the same type of utility, the question of their consistency arises. I explore such consistency issues concerning hybrids, too.

This appendix answers a representative sample of questions about consistency but does not answer all such questions. Some unaddressed questions may be answered using the techniques illustrated, and are left as exercises for the reader.

A.1. APPROACHES TO CONSISTENCY

The methods of utility analysis are not blatantly inconsistent. They do not, for example, explicitly claim that $U(o) = 1$ and that $U(o) = 2$, or that $U(o) = P(s)$ and that $U(o) = P(\sim s)$ in cases where $P(s) \neq P(\sim s)$. Still the methods may be implicitly inconsistent. The methods of analyzing $U(o)$ give its value in terms of other utilities and in some cases probabilities. The methods impose constraints on utility assignments as the basic probability laws impose constraints on probability assignments. For example, expected utility analysis places constraints on the values of options' utilities and conditional utilities for option-state pairs. Do the constraints conflict in some cases because of defects in the principle of pros and cons or its application to a method of utility analysis? I want to show the consistency of the methods of utility analysis by showing that it is possible to meet all the constraints they impose wherever they are applicable. How should I proceed?

The main obstacle to finding independent argumentation for consistency is that comprehensive and intrinsic utility, being primitives in my theory, do not have definitions that generate a canonical method of computing utilities. Hence I cannot show consistency the way one shows, for example, the consistency of the various methods of adding a set of numbers by deriving association and commutation laws from basic definitions of numbers and addition. I have to construct an alternative procedure.

One way to argue for the consistency of my applications of the principle of pros and cons is to adopt a canonical specification of pros and cons, and an associated canonical method of computing utilities, and then argue that every application of the principle introduces a method of computing utilities equivalent to the canonical method. The applications introduce methods of adding the canonical pros and cons that differ only in grouping and order, the argument claims, and so are equivalent by the association and commutation laws of addition.[1]

I adopt this approach to consistency. I specify a canonical method of computing utilities, which implies a canonical category of pros and cons. Then I use the canonical method to establish agreement between the various forms of utility analysis. I reduce them to a canonical computation of utilities. The reduction requires canonical methods of obtaining all the input for the various forms of utility analysis, includ-

---

[1] Compare material on crosscutting additive separability, for example, Broome (1991: chap. 4). This material treats the representation of utilities from various dimensions by a single utility function for cells. A representation theorem states conditions under which a utility function for cells consistently yields utilities of combinations of cells if utilities of cells are added according to the methods of utility analysis for the dimensions.

ing probabilities for expected utility analysis. So the canonical methods are, first, a canonical form of utility analysis and, second, a canonical way of obtaining inputs for the other forms of utility analysis.

The consistency of two methods of analyzing $U(o)$ is easy to establish if the two methods have independent input. The input for one method may be adjusted to give the same value for $U(o)$ as the other method. The possibility of inconsistency arises only if the two methods use the same input, or interderivable input, so that the input for the two methods cannot be adjusted independently to yield the same value for $U(o)$. Intrinsic utility analysis cannot be inconsistent with expected utility analysis, for example, unless the input for one analysis has implications about the input for the other. To provide for the possibility of inconsistency, I assume that the input for all the methods of utility analysis is derivable from probabilities of worlds and intrinsic utilities of objects of basic intrinsic attitudes.

Is there an alternative method of showing consistency that dispenses with canonical methods of computing utilities? One may argue for the consistency of applications of the principle of pros and cons by assuming the consistency of utilities derived from some partitions of pros and cons and then arguing for the consistency of utilities derived from all partitions of pros and cons. For example, one might assume that if R is a multielement partition of pros and cons, or reasons, and R' is a finer partition obtained by partitioning some elements of R, then utilities derived from R and R' are consistent. It follows from this assumption that for any two partitions of reasons, because there is a subsuming partition whose elements are contained by elements of each of the two original partitions, the utilities derived from the two partitions agree with the utility derived from the subsuming partition, and so agree with each other. The consistency of utilities obtained from each pair of partitions then entails the consistency of the utilities obtained from all partitions.

Appearances notwithstanding, this method of showing consistency is not more economical than my approach. It requires showing that the various forms of utility analysis all partition pros and cons. The text does not provide the requisite demonstrations because it does not examine the nature of pros and cons. It rests on the intuition that the forms of utility analysis partition pros and cons. To provide independent verification that the forms of utility analysis partition pros and cons, I have to specify canonical pros and cons and argue that the forms of utility analysis partition them. This requires specification of a canonical method of computing utilities. So the alternative approach does not really dispense with a specification of a canonical method of computing utilities. Furthermore, it assumes the consistency of some forms of utility analysis. I prefer a more fundamental approach that does not begin with consistency assumptions.

Because this appendix argues for consistency only, not correctness, I am free, technically speaking, to adopt any method of computing utilities as the canonical method. It does not have to be a method introduced in the text. It does not even have to be correct. Consistency requires only the possibility of agreement between the forms of utility analysis. For consistency, it suffices to show that there is some method of computing utilities such that, given it, the forms of utility analysis agree. Now obviously agreement of the forms of utility analysis can be achieved by adopting a canonical method of computation that makes all utilities equal zero; all probability-utility products and their sums are then zero. But such a canonical method lacks realism. I want to show consistency given a canonical method that recognizes a realistic range of probability and utility assignments. I want a canonical method that countenances nontrivial utility assignments and nontrivial probability assignments. Given this stipulation, the canonical method must for the sake of practicality aim at correctness so that the methods of utility analysis, which aim at correctness, agree with it. Thus I select as the canonical method a hybrid application of forms of utility analysis already supported in the text.

The objective is to show that given the canonical method – and some way of computing the input for the forms of utility analysis – the forms of utility analysis agree so that utilities computable by two forms of analysis have the same value according to those forms of analysis. However, because my forms of utility analysis are advanced as a group only for ideal cases, I need to show agreement only in ideal cases. An ideal agent is cognitively perfect and has a fully rational state of mind and, when in ideal conditions, assigns probabilities and utilities sufficient for conducting utility analyses. The idealization ensures consistency of the probabilities and utilities forming the input for the forms of utility analysis, so that if two analyses yield different values for a utility, the difference stems from the analyses' inconsistency, and not from some defect in the agent.[2]

---

[2] Weirich (1977: chap. 1) shows that compliance with the standard probability laws is necessary for the consistency of expected utility analysis across all ways of

The text provides no definition of rationality. It advances forms of utility analysis as rationality constraints, not as a definition of rationality. In the argument for consistency, I am therefore free, technically speaking, to take liberties with the interpretation of rationality. To show consistency, it suffices to show agreement of the forms of utility analysis in all rational cases, given any interpretation of rationality. Agreement may be achieved easily by stipulating that rationality requires compliance with all forms of utility analysis. In this case no assignment of degrees of belief and desire that generates an inconsistency is rational; ideal agents do not make such an assignment. For example, if two forms of analysis claim that $U(o) = P(s)$ and that $U(o) = P(\sim s)$, then, given an interpretation of rationality that stipulates compliance with those forms of analysis, rationality requires that $P(s) = P(\sim s)$. If the forms of utility analysis are stipulated as part of the interpretation of rationality, then inconsistency is impossible in rational cases.

This type of demonstration of consistency is unsatisfactory. I want to demonstrate consistency given plausible, liberal accounts of rationality. I want to head off objections that the forms of utility analysis are inconsistent, given nonrestrictive views of rationality. So I adopt a tolerant interpretation of rationality requirements on the input for the canonical methods of computing probabilities and utilities.

Although the canonical methods of obtaining probabilities and utilities need not be foundational, I adopt methods that are foundational. That is, their input are reasons as fine-grained as any. I assume that in ideal cases there is a finest level of reasons, namely, probabilities of possible worlds and intrinsic utilities of objects of basic intrinsic attitudes (BITs). I have already assumed that these probabilities and utilities yield the input for the forms of utility analysis. Now I assume that no reasons are finer-grained than these, and I take them as canonical reasons. I also assume, as in the text, a finite number of BITs and then, after trimming irrelevancies, a finite number of possible worlds.

As the canonical method of computing utility, I take a combination of expected and intrinsic utility analyses, specifically, expected utility analysis with respect to possible worlds and intrinsic utility analysis of

partitioning states. This result provides an argument for the probability laws, granting the consistency of expected utility analysis. Here I work in the opposite direction. I show that, given rationality constraints on the input for expected utility analysis, the analysis is consistent for all partitions of states.

worlds in terms of BITs they realize. The canonical method of computing utility is thus:

$$U(o) = \Sigma_i P(w_i \text{ given } o)\Sigma_{j \in \{ji\}} IU(\text{BIT}_j),$$

where w_i ranges over possible worlds and $\{ji\}$ indexes the objects of the BITs realized in w_i. In other words, I use the rule that the utility of an option is a probability-weighted average of the sums of the intrinsic utilities of the sets of BITs that might be realized if the option were realized (see Section 5.2.2). I show that all methods of calculating the utility of an option agree with this method and so are consistent with each other. It would be simple to extend the consistency results to all propositions, but I focus on options because the primary topic is the application of utility analysis to decision problems.

I also need canonical methods of obtaining input for all forms of utility analysis from probabilities of worlds and intrinsic utilities of objects of BITs. For instance, I need a canonical method of computing utilities of the form $U(o$ given $(s$ if $o))$ for expected utility analysis. As canonical methods of obtaining input, I adopt familiar methods, presented in the course of the consistency proofs. For instance, I derive probabilities of states from probabilities of worlds in the usual way.

My forms of utility analysis use two factors, namely, beliefs and desires. In ideal cases a fine-grained analysis of beliefs yields a probability distribution over possible worlds. A fine-grained analysis of desires yields utilities for objects of BITs. Hence, a fine-grained utility analysis uses expected utility analysis with possible worlds as states and intrinsic utility analysis with singletons of BITs as the sets of BITs with respect to which the analysis is conducted. When the analyses of beliefs and desires are combined, the fundamental unit of analysis is the intrinsic utility of a BIT's realization in a possible world, weighted by the probability of the possible world. Every analysis of $U(o)$ is reducible to operations on pros and cons of this type. The various forms of utility analysis differ only in their ways of adding these canonical pros and cons. No matter how the addition takes place, the separability of these pros and cons guarantees that the results are the same.

The canonical methods of computing probabilities and utilities do not assume any substantive constraints on the probabilities of worlds, the intrinsic utilities of objects of BITs, or the nearness relations involved in conditional probabilities and utilities. For a single a posteriori proposition, any probability or utility value is possible. The methods do, however, assume basic coherence constraints. In particular, following the idealization about rationality, they assume utility assignments that conform to the canonical analysis and so, for instance, make the comprehensive utility of a world equal the sum of the intrinsic utilities of the objects of the BITs realized there. In addition, they assume intrinsic utility assignments meeting the basic coherence requirements for BITs so that, for example, two contradictory propositions do not both receive positive intrinsic utilities.

Using the canonical method of utility analysis to show the consistency of the other forms of utility analysis also shows their correctness given the canonical method's correctness, if we assume that the demonstration of consistency uses an interpretation of rationality sufficiently liberal to encompass all genuinely rational cases. Hence the argument for the consistency of the forms of utility analysis may be strengthened to obtain an argument for their correctness by arguing that the canonical method is correct. Because Chapter 5 supports the canonical method, the argument for the consistency of my forms of utility analysis is a supplementary argument for their correctness. Although this supplementary argument is closely allied to the text's argument for their correctness, it is independent. For example, the text supports expected utility analysis independently of intrinsic utility analysis, whereas the supplementary argument derives expected utility analysis from a type of intrinsic utility analysis.

A.2. EXPECTED UTILITY ANALYSIS WITH VARIOUS PARTITIONS

Sobel (1994: chap. 1) uses expected utility analysis with respect to worlds, which he calls "world Bayesianism," as the canonical method of calculating utility. He assumes certain methods of using the input for world Bayesianism to obtain the input for other expected utility analyses and then provides "partition theorems" for those analyses, theorems establishing their consistency with world Bayesianism by reducing them to world Bayesianism (chap. 9). Specifically, a partition theorem shows that the expected utility of an option with respect to any partition equals its expected utility with respect to the ultimate partition, the partition of possible worlds. This section proves this type of theorem for my version of expected utility analysis. It reduces all expected utility analyses of a proposition's utility to one using the partition of worlds, adopting expected conditional utility analysis using

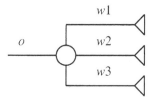

Figure A.1. Worlds

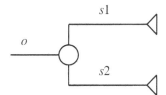

Figure A.2. States

worlds as the canonical way of obtaining input for the other analyses. However, I do not stop after establishing a partition theorem because expected utility analysis with respect to worlds is not canonical for me. Because I adopt as canonical a combination of expected and intrinsic utility analyses, the next section reduces expected utility analysis using worlds to that hybrid form of utility analysis.

This section's strategy can be illustrated with expected utility trees. Imagine that there are three possible worlds w_1, w_2, and w_3. Then with respect to worlds $U(o)$ has the expected utility tree in Figure A.1. With respect to a coarser partition of states, s_1 and s_2, $U(o)$ also has the expected utility tree in Figure A.2. To show that the second expected utility tree yields the same value for $U(o)$ as the first, expand the terminal nodes of the second tree using expected conditional utility analysis until reaching worlds. The expansion provides the input utilities for the original tree using states. Assuming that s_1 is the disjunction w_1 or w_2, and s_2 is w_3, it generates the tree in Figure A.3. Then show that $U(o)$ according to the expanded tree is the same as its expected utility according to the original tree using the partition of worlds. This establishes that the option's expected utility according to the states is the same as its expected utility according to the worlds, because, by the canon, the option's expected utility according to the states is the same as its expected utility according to the expanded tree.

251

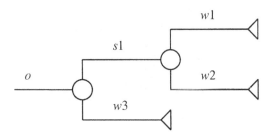

Figure A.3. State Expansion

Using the set of possible worlds to form a partition of states, Section 4.2's version of expected utility analysis yields that $U(o) = \Sigma_i P(w_i$ given $o)U(o$ given $(w_i$ if $o))$. Here it is assumed that w_i ranges over worlds such that it is possible that w_i if o. One may simplify the formula because an option helps identify an outcome only when a state is incomplete; the option's job is to remedy the state's incompleteness. Because there is nothing incomplete about a possible world, the outcome of o given $(w_i$ if $o)$ is the same as the outcome if w_i. Hence $U(o$ given $(w_i$ if $o)) = U(w_i)$, and therefore $U(o) = \Sigma_i P(w_i$ given $o)U(w_i)$. The restriction on w_i's range may be lifted because utilities are no longer conditional.

Next I show that $U(o)$ equals the same sum when analyzed with respect to any partition of states. Take the partition of states $\{s_j\}$. According to expected utility analysis, $U(o) = \Sigma_j P(s_j$ given $o)U(o$ given $(s_j$ if $o))$. To expand at terminal nodes the tree representing this analysis, for each j form the partition of s_j into worlds $\{w_{kj}\}$. By expected conditional utility analysis, $U(o$ given $(s_j$ if $o)) = \Sigma_{kj} P_o(w_{kj}$ given $s_j)U(o$ given $([w_{kj}$ & $s_j]$ if $o))$, where w_{kj} ranges over worlds of the partition for s_j. Given that $P_o(w_{kj}$ given $s_j) = P_o(w_{kj}/s_j)$ according to the interpretation of $P_o(w_{kj}$ given $s_j)$ Section 4. A adopts, $U(o$ given $(s_j$ if $o)) = \Sigma_{kj}[P_o(w_{kj}$ & $s_j)/P_o(s_j)]U(o$ given $([w_{kj}$ & $s_j]$ if $o))$. Because w_{kj} & s_j is logically equivalent to w_{kj}, this sum reduces to $\Sigma_{kj}[P_o(w_{kj})/P_o(s_j)]U(o$ given $(w_{kj}$ if $o))$. It reduces further to $\Sigma_{kj}[P_o(w_{kj})/P_o(s_j)]U(w_{kj})$ because $U(o$ given $(w_{kj}$ if $o)) = U(w_{kj})$, as the previous paragraph explains for w_i. Substituting for $U(o$ given $(s_j$ if $o))$ in the formula for $U(o)$ computed with respect to $\{s_j\}$ generates the identity $U(o) = \Sigma_j P(s_j$ given $o)\Sigma_{kj}[P_o(w_{kj})/P_o(s_j)]U(w_{kj})$, or $\Sigma_j P_o(s_j)\Sigma_{kj}[P_o(w_{kj})/P_o(s_j)]U(w_{kj})$. By algebra, this sum equals $\Sigma_j\Sigma_{kj}P_o(w_{kj})U(w_{kj})$, which in turn equals $\Sigma_i P_o(w_i)U(w_i)$, that is, $\Sigma_i P(w_i$ given $o)U(w_i)$, because as j and k vary they reach all worlds, each exactly once. Thus an expected utility analysis of

252

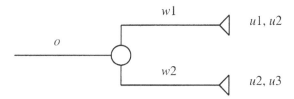

Figure A.4. Expected and Intrinsic Utility Analyses

$U(o)$ using the partition $\{s_j\}$ and splitting $U(o$ given $(s_j$ if $o))$ with a partition of worlds is equivalent to an expected utility analysis of $U(o)$ using a partition of worlds. Both give the same results.

A.3. EXPECTED AND INTRINSIC UTILITY ANALYSES

In light of the previous section's results, one may show that expected utility analysis is consistent with intrinsic utility analysis by showing that expected utility analysis using worlds is consistent with intrinsic utility analysis. I show the latter by showing that both an expected utility analysis of $U(o)$ using worlds and an intrinsic utility analysis of $U(o)$ are reducible to a canonical utility analysis of $U(o)$. This section treats Section 5.2.3's form of intrinsic utility analysis using intrinsic utilities of objects of basic intrinsic attitudes (BITs). The next section treats Section 5.3's form of intrinsic utility analysis using partial utility functions.

To illustrate the consistency of expected and intrinsic utility analysis, consider the simple two-world expected utility analysis in Figure A.4. I assume that the utility of the world at the top terminal node is the sum of the intrinsic utilities of the BITs realized there, u_1 and u_2. Similarly for the world at the bottom terminal node. This additivity follows from the canonical method of obtaining input for expected utility analyses. To simplify the illustration, I also assume that the worlds are probabilistically independent of the option (it follows that the option is not part of a world after trimming irrelevant detail and so is not an intrinsic attitude's object). By expected utility analysis, $U(o)$ $= P(w_1)(u_1 + u_2) + P(w_2)(u_2 + u_3)$. But also by intrinsic utility analysis, $U(o) = P(w_1)u_1 + [P(w_1) + P(w_2)]u_2 + P(w_2)u_3$. In this case the two analyses plainly yield the same result.

The reduction of expected and intrinsic utility analyses to the canonical form of utility analysis requires a canonical way of obtaining their input from its input: probabilities of worlds and intrinsic utilities of objects of BITs. The canonical method of obtaining the comprehensive utility of a world says that $U(w) = \Sigma_i IU(\mathrm{BIT}_i)$, where BIT_i ranges over the objects of BITs realized in w. The canonical method of obtaining the probability of a BIT's object from probabilities of worlds says that $P(\mathrm{BIT}_j$ given $o)$ is the sum of the probabilities of the worlds where BIT_j is realized given o. That is, $P(\mathrm{BIT}_j$ given $o) = \Sigma_{i \in \{ij\}} P(w_i$ given $o)$, where $\{ij\}$ indexes the worlds in which BIT_j is realized given o.

By expected utility analysis using worlds as states, $U(o) = \Sigma_i P(w_i$ given $o)U(o$ given $(w_i$ if $o))$. Because $U(o$ given $(w_i$ if $o)) = U(w_i)$, as Section A.2 explains, $U(o) = \Sigma_i P(w_i$ given $o)U(w_i)$. Also, a possible world's comprehensive utility is the same as its intrinsic utility; a world is a full specification of its outcome if it were realized and so it entails everything that affects its utility. Thus $U(w_i) = IU(w_i)$. Moreover, the intrinsic utility of a possible world is the sum of the intrinsic utilities of the objects of BITs realized there, according to the canonical method of obtaining the input for expected utility analysis. Hence, $U(o) = \Sigma_i P(w_i$ given $o)\Sigma_{j \in \{ji\}} IU(\mathrm{BIT}_j)$, where $\{ji\}$ indexes the objects of BITs realized in w_i. This sum is the same given by the canonical analysis of $U(o)$.

On the other hand, by intrinsic utility analysis, $U(o) = \Sigma_j P(\mathrm{BIT}_j$ given $o)IU(\mathrm{BIT}_j)$, where BIT_j ranges over objects of BITs. By the canonical method of obtaining input, $P(\mathrm{BIT}_j$ given $o) = \Sigma_{i \in \{ij\}} P(w_i$ given $o)$, where $\{ij\}$ indexes the worlds in which BIT_j is realized. So $U(o) = \Sigma_j [\Sigma_{i \in \{ij\}} P(w_i$ given $o)]IU(\mathrm{BIT}_j) = \Sigma_j \Sigma_{i \in \{ij\}} P(w_i$ given $o)IU(\mathrm{BIT}_j) = \Sigma_{j,i \in \{ij\}} P(w_i$ given $o)IU(\mathrm{BIT}_j)$. Next, notice that for any i and $j \in \{ji\}$, because BIT_j is realized in w_i, $i \in \{ij\}$. Conversely, for any j and $i \in \{ij\}$, because BIT_j is realized in w_i, $j \in \{ji\}$. So $\Sigma_{j,i \in \{ij\}} P(w_i$ given $o)IU(\mathrm{BIT}_j) = \Sigma_{i,j \in \{ji\}} P(w_i$ given $o)IU(\mathrm{BIT}_j)$. By algebra, $\Sigma_{i,j \in \{ji\}} P(w_i$ given $o)IU(\mathrm{BIT}_j) = \Sigma_i \Sigma_{j \in \{ji\}} P(w_i$ given $o)IU(\mathrm{BIT}_j) = \Sigma_i P(w_i$ given $o)\Sigma_{j \in \{ji\}} IU(\mathrm{BIT}_j)$. Therefore intrinsic utility analysis also yields the same value for $U(o)$ as does canonical utility analysis.

Because both expected and intrinsic utility analyses of an option's utility reduce to the canonical utility analysis of the option's utility, the two forms of utility analysis are consistent. An analogous proof shows that expected and intrinsic conditional utility analyses are also consistent.

A.4. PARTIAL AND CANONICAL UTILITY ANALYSES

To show the consistency of partial utility analysis with the other forms of utility analysis, this section reduces partial utility analysis to the canonical form of utility analysis. Partial utility analysis, generalized in Section 5.A.2, uses a partition of BITs to define a set of partial utility functions U_j. To begin, I treat Section 5.3's special case, in which the members of the partition are singletons of BITs and U_j is taken with respect to a single BIT designated BIT$_j$.

The reduction of partial utility analysis must specify a canonical method of calculating $U_j(o)$, the input for partial utility analysis. Section 5.3 supported the formula $U_j(o) = P(\text{BIT}_j \text{given } o)IU(\text{BIT}_j)$, but this formula needs support from a canonical formula that uses probabilities of worlds in place of $P(\text{BIT}_j \text{given } o)$. As the canonical formula, I use the equivalent identity $U_j(o) = \Sigma_{i \in \{ij\}} P(w_i \text{ given } o)IU(\text{BIT}_j)$, where $\{ij\}$ indexes the worlds where BIT$_j$ is realized. Having given Section 5.3's formula canonical roots, a reduction of partial utility analysis may employ it when convenient.

According to partial utility analysis in Section 5.3, $U(o) = \Sigma_j U_j(o)$, where j ranges over BITs. Given that $U_j(o) = P(\text{BIT}_j \text{given } o)IU(\text{BIT}_j)$, it follows that $U(o) = \Sigma_j P(\text{BIT}_j \text{ given } o)IU(\text{BIT}_j)$. Also, according to intrinsic utility analysis as presented in Section 5.2.3, $U(o) = \Sigma_j P(\text{BIT}_j$ given $o)IU(\text{BIT}_j)$. Because, as Section A.3 establishes, the intrinsic utility analysis reduces to the canonical utility analysis, the partial utility analysis also reduces to the canonical analysis.

To address Section 5.A.2's general form of partial utility analysis, suppose that U_j is computed with respect to a set of BITs. Then the canonical method of obtaining input for the partial utility analysis is $U_j(o) = \Sigma_{k \in \{j\}} \Sigma_{i \in \{k\}} P(w_i \text{ given } o)IU(\text{BIT}_k)$, where $\{j\}$ indexes the BITs relevant to U_j and $\{k\}$ indexes the worlds where BIT$_k$ is realized. This formula may be used to reduce $U(o)$'s partial utility analysis to its canonical utility analysis. The details are left as an exercise.

A.5. HYBRID FORMS OF UTILITY ANALYSIS

Although the reduction of each form of utility analysis to the canonical form entails the consistency of a hybrid combining expected and intrinsic utility analyses with a hybrid combining expected and partial utility analyses, let me independently verify the consistency of these two hybrid analyses.

As a preliminary, I consider the utility of a possible world w with respect to BIT$_j$, or $U_j(w)$. There is no uncertainty concerning the outcome of w. If BIT$_j$ is realized in w, then $U_j(w) = IU(\text{BIT}_j)$. If it is not, then $U_j(w) = 0$, because all that matters to U_j is BIT_j's realization, as Section 5.3 explains.

Now consider a combination of expected and intrinsic utility analyses applied with respect to a partition of worlds: $U(o) = \Sigma_i P(w_i$ given $o)IU(w_i)$, where w_i ranges over possible worlds. If we apply the canonical method of obtaining $IU(w_i)$, $U(o) = \Sigma_i P(w_i$ given $o)\Sigma_{j\in\{ji\}}IU(BIT_j)$, where w_i ranges over possible worlds and $\{ji\}$ indexes the objects of BITs realized in w_i. Then by algebra, $U(o) = \Sigma_{j,i\in\{ij\}}P(w_i$ given $o)IU(BIT_j)$, where $\{ij\}$ indexes the worlds in which BIT_j is realized (see Section A.3). Substituting $U_j(w_i)$ for $IU(BIT_j)$, generates the identity $U(o) = \Sigma_{j,i\in\{ij\}}P(w_i$ given $o)U_j(w_i)$.

On the other hand, according to a combination of partial and expected utility analyses with respect to worlds, $U(o) = \Sigma_j\Sigma_i P(w_i$ given $o)U_j(o$ given $(w_i$ if $o))$. This yields the identity $U(o) = \Sigma_j\Sigma_i P(w_i$ given $o)U_j(w_i)$ because, as Section A.2 observes, $U_j(o$ given $(w_i$ if $o)) = U_j(w_i)$. By algebra, then, $U(o) = \Sigma_{j,i}P(w_i$ given $o)U_j(w_i)$. Now if BIT_j is not realized in w_i, then $U_j(w_i) = 0$. When the last sum omits summands equal to zero, the last formula for $U(o)$ becomes the formula at the end of the previous paragraph. Thus the two hybrid utility analyses agree.

A.6. EXPECTED, INTRINSIC, AND GROUP UTILITY ANALYSES

Section 6.3 defines a group's utility assignment in terms of its members' utility assignments: $GU(o) = \Sigma_i e_i U_i(o)$, where e_i is the economic power of the ith member of the group and U_i is his utility assignment. Application of the definition is a group utility analysis of $GU(o)$. To support the definition, this section establishes two consistency results. First, it shows that expected utility analysis applied to group utility is consistent with expected utility analysis applied to members' utilities, provided that the members have a common probability assignment that forms the group's probability assignment. Second, it shows that given a group probability assignment, however obtained, expected and intrinsic utility analyses yield consistent results when applied to group utility, that is, when groups are treated as agents with probability assignments and basic intrinsic attitudes as well as group utility assignments.

The first consistency result establishes that if the members of a group have the same probability assignment, then two methods of computing $GU(o)$ yield the same value. The first method applies expected

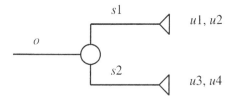

Figure A.5. Expected and Group Utility Analyses

utility analysis to $GU(o)$ and then applies the definition of group utility to o's possible outcomes. The second method applies the definition of group utility to $GU(o)$ and then applies expected utility analysis to the members' utility assignments for o.

To compare the two methods, consider the utility tree in Figure A.5 for a two-person group. Here the first element in the pair of utilities at a terminal node is the utility of the outcome for the first member, and the second element is the utility of the outcome for the second member. Let P be the common probability assignment of the two members. It is taken as the group's probability assignment. According to expected utility analysis for groups, if we assume that the states are independent of the option, $GU(o) = P(s_1)GU(o \text{ given } s_1) + P(s_2)GU(o \text{ given } s_2)$. If we apply the definition of group utility to the outcomes' group utilities, this sum equals $P(s_1)[e_1u_1 + e_2u_2] + P(s_2)[e_1u_3 + e_2u_4]$. On the other hand, according to the definition of group utility, $GU(o) = e_1U_1(o) + e_2U_2(o)$. If we apply expected utility analysis to members' utilities, this sum equals $e_1[P(s_1)u_1 + P(s_2)u_3] + e_2[P(s_1)u_2 + P(s_2)u_4]$. Obviously, the two ways of computing $GU(o)$ are equivalent.[3]

Now I show that a common probability assignment entails that the two methods of computing group utility agree in all cases. Suppose that P is the members' common probability assignment and so the group's probability assignment. Then, according to expected utility analysis applied to an option's group utility, $GU(o) = \Sigma_j P(s_j \text{ given } o)GU(o \text{ given }$

---

[3] Section 6.4.1 observes that compatibility of the two methods of calculating group utility entails a common probability assignment. In the example it is plain that a common probability assignment is necessary if the two methods of computing group utility are to be consistent. In the absence of a common probability assignment, an option's group utility is computed from the option's utilities for members by applying expected utility analysis to members' utilities using distinct members' probability assignments. Also, some unspecified group probability function is used to compute the option's group utility using expected utility analysis. In this case the two ways of calculating the option's group utility may disagree.

257

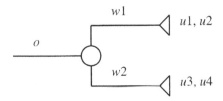

Figure A.6. Expected, Intrinsic, and Group Utility Analyses

(s_j if o)). Applying the definition of group utility to outcomes yields the identity $GU(o) = \Sigma_j P(s_j$ given $o)\Sigma_i e_i U_i(o$ given (s_j if o)). By algebra, the sum on the right equals $\Sigma_i e_i \Sigma_j P(s_j$ given $o)U_i$ (o given (s_j if o)). By expected utility analysis for members' utilities, this sum is $\Sigma_i e_i U_i(o)$, which equals $GU(o)$ according to the definition of an option's group utility. Hence expected utility analysis for groups is consistent with expected utility analysis for members.

Next, consider the consistency of expected and intrinsic utility analyses applied to group utility. The two analyses assume a group probability assignment P but do not assume anything about members' probability assignments because they do not apply expected utility analysis to members' utilities. Figure A.6's decision tree for a two-person group illustrates the two methods of computing group utility. The branches indicate possible worlds. The utility on the left at a terminal node is the utility of the associated world for the first group member, and the utility on the right is the utility of the associated world for the second group member. The utility of a world for an agent accrues from realization of a single basic intrinsic attitude (BIT), whose object is distinct from the object of any other BIT of any group member. Section 5.2.3's version of intrinsic utility analysis applied to an option's group utility says that $GU(o) = \Sigma_j P(\text{BIT}_j$ given $o)GIU(\text{BIT}_j)$. Here $GIU(\text{BIT}_j)$ is the strength of the group's intrinsic attitude toward realization of BIT_j, a basic intrinsic attitude of a member of the group. According to Section 6.3's definition of GIU applied to outcomes, the equation implies that $GU(o) = \Sigma_j P(\text{BIT}_j$ given $o)\Sigma_i e_i IU_i(\text{BIT}_j)$, where BIT_j ranges over objects of BITs. Because in the example each BIT is realized in only one world, and pertains to only one group member, and because for worlds comprehensive utility equals intrinsic utility, the latter sum equals $[P(w_1$ given $o)e_1 u_1 + P(w_1$ given $o)e_2 u_2] + [P(w_2$ given $o)e_1 u_3 + P(w_2$ given $o)e_2 u_4]$.

Now I apply expected utility analysis to $GU(o)$ using a partition of worlds. According to the analysis, $GU(o) = \Sigma_k P(w_k$ given $o)GU(w_k)$

because $GU(o$ given $(w_k$ if $o)) = GU(w_k)$. See Section A.2. Applying the definition of GU to outcomes, the sum equals $\Sigma_k P(w_k$ given $o)\Sigma_i e_i U_i(w_k)$. Because in the example each world realizes exactly one of each agent's BITs, which provides the utility of the world for him, the latter sum equals $P(w_1$ given $o)(e_1 u_1 + e_2 u_2) + P(w_2$ given $o)(e_1 u_3 + e_2 u_4)$. It is easy to see that this method of calculating $GU(o)$ yields the same result as the first method.

To prove agreement in general, I consider in turn each of the two formulas for group utility. According to intrinsic utility analysis, $GU(o) = \Sigma_j P(\text{BIT}_j$ given $o)GIU(\text{BIT}_j) = \Sigma_j P(\text{BIT}_j$ given $o)\Sigma_i e_i IU_i(\text{BIT}_j)$. Because $P(\text{BIT}_j$ given $o)$ equals $\Sigma_{k\in\{kj\}} P(w_k$ given $o)$, where $\{kj\}$ indexes the worlds in which BIT_j is realized, $GU(o) = \Sigma_j\Sigma_{k\in\{kj\}} P(w_k$ given $o)\Sigma_i e_i IU_i(\text{BIT}_j)$. By algebra, the latter equals $\Sigma_{j,i,k\in\{kj\}} P(w_k$ given $o)e_i IU_i(\text{BIT}_j)$, which equals $\Sigma_{k,i,j\in\{jk\}} P(w_k$ given $o)e_i IU_i(\text{BIT}_j)$. See Section A.3.

Next, according to expected and then group utility analysis, $GU(o) = \Sigma_k P(w_k$ given $o)GU(w_k) = \Sigma_k P(w_k$ given $o)\Sigma_i e_i U_i(w_k)$. By intrinsic utility analysis applied to members' utilities, the last sum equals $\Sigma_k P(w_k$ given $o)\Sigma_i e_i \Sigma_{j\in\{jk\}} IU_i$ (BIT_j), where $\{jk\}$ indexes the BITs realized in w_k. By algebra this equals $\Sigma_{k,i,j\in\{jk\}} P(w_k$ given $o)e_i IU_i(\text{BIT}_j)$.

From the conclusions of the two preceding paragraphs, it follows that expected and intrinsic utility analyses agree when applied to group utility. Both methods of computation say that $GU(o) = \Sigma_{k,i,j\in\{jk\}} P(w_k$ given $o)e_i IU_i(\text{BIT}_j)$.

The foregoing proofs of consistency involve shortcuts. A standard proof reduces each of two methods of calculating $GU(o)$ to the canonical method, whereby $GU(o) = \Sigma_i e_i U_i(o) = \Sigma_i e_i \Sigma_{k,j\in\{jk\}} P_i(w_k$ given $o)IU_i(\text{BIT}_j)$. This is the method that computes members' utilities from their probability assignments to worlds and their intrinsic utility assignments to objects of BITs. The additional steps needed for reductions to the canonical method are straightforward. For the consistency result concerning intrinsic and expected utility analyses, the standard proof's additional reductive steps assume that the group's members have a common probability assignment P that forms the group's probability assignment. Then the canonical formula may be rewritten as $GU(o) = \Sigma_i e_i \Sigma_{k,j\in\{jk\}} P(w_k$ given $o)IU_i(\text{BIT}_j)$. By algebra, this is equivalent to $GU(o) = \Sigma_{k,i,j\in\{jk\}} P(w_k$ given $o)e_i IU_i(\text{BIT}_j)$, which as demonstrated earlier, is the result of both intrinsic and expected utility analyses of $GU(o)$.

259

References

Allais, M. 1953. "Le comportement de l'homme rationnel devant le risque." *Econometrica* 21:503–46. Translated as "Foundations of a Positive Theory of Choice Involving Risk and a Criticism of the Postulates and Axioms of the American School." In Maurice Allais and Ole Hagen, eds., *Expected Utility Hypothesis and the Allais Paradox*, pp. 27–145. Dordrecht: Reidel, 1979.

Allais, M., and O. Hagen, eds. 1993. *Cardinalism: A Fundamental Approach.* Dordrecht: Kluwer.

Armendt, B. 1986. "A Foundation for Causal Decision Theory." *Topoi* 5:3–19.

——— 1988. "Conditional Preference and Causal Expected Utility." In W. Harper and B. Skyrms, eds., *Causation in Decision, Belief Change, and Statistics*, 2:1–24. Dordrecht: Kluwer.

Arrow, K. 1951. *Social Choice and Individual Values.* New Haven: Yale University Press.

——— 1965. *Aspects of the Theory of Risk-Bearing.* Helsinki: Yrjö Jahnssonin Säätiö.

——— 1970. *Essays in the Theory of Risk-Bearing.* Amsterdam: North-Holland.

Audi, R, ed. 1995. *The Cambridge Dictionary of Philosophy.* Cambridge: Cambridge University Press.

Aumann, R. 1976. "Agreeing to Disagree." *Annals of Statistics* 4:1236–9.

Barry, B. 1976. "Power: An Economic Analysis." In B. Barry, ed., *Power and Political Theory*, pp. 67–101. New York: Wiley.

Beebee, H., and D. Papineau. 1997. "Probability as a Guide to Life." *Journal of Philosophy* 94:217–43.

Bell, D., and H. Raiffa. 1988. "Marginal Value and Intrinsic Risk Aversion." In D. Bell, H. Raiffa, and A. Tversky, eds., *Decision Making*, pp. 384–97. Cambridge: Cambridge University Press.

Binmore, K. 1994. *Game Theory and the Social Contract.* Vol. 1, *Playing Fair.* Cambridge, MA: MIT Press.

——— 1998. *Game Theory and the Social Contract.* Vol. 2, *Just Playing.* Cambridge, MA: MIT Press.

Binmore, K., and P. Dasgupta. 1987. *The Economics of Bargaining.* Oxford: Blackwell.

Bordley, R. 1986. "Review Essay: Bayesian Group Decision Theory." In B. Grofman and G. Owen, eds., *Information Pooling and Group Decision Making*, pp. 49–68. Greenwich, CT: JAI Press.

Bowie, N., and R. Freeman, eds. 1992. *Ethics and Agency Theory*. New York: Oxford University Press.

Brady, M. 1993. "J. M. Keynes's Theoretical Approach to Decision-Making under Conditions of Risk and Uncertainty." *British Journal for the Philosophy of Science* 44:357–76.

Brandenburger, A., and B. Nalebuff. 1996. *Co-opetition*. New York: Doubleday.

Brandt, R. 1979. *A Theory of the Good and the Right*. Oxford: Oxford University Press.

Bratman, M. 1987. *Intention, Plans, and Practical Reason*. Cambridge, MA: Harvard University Press.

Brennan, T. 1989. "A Methodological Assessment of Multiple Utility Frameworks." *Economics and Philosophy* 5:189–208.

Broome, J. 1987. "Utilitarianism and Expected Utility." *Journal of Philosophy* 84:405–22.

1991. *Weighing Goods*. Oxford: Blackwell.

1999. *Ethics out of Economics*. Cambridge: Cambridge University Press.

Buchanan, A., and D. Brock. 1989. *Deciding for Others*. Cambridge: Cambridge University Press.

Bueno de Mesquita, Bruce. 1981. *The War Trap*. New Haven: Yale University Press.

Campbell, R., and L. Sowden. 1985. *Paradoxes of Rationality and Cooperation*. Vancouver: University of British Columbia Press.

Carnap, R. 1966. *Philosophical Foundations of Physics*. New York: Basic Books.

Chalmers, D. 1996. *The Conscious Mind*. New York: Oxford University Press.

Chang, R., ed. 1997. *Incommensurability, Incomparability, and Practical Reason*. Cambridge, MA: Harvard University Press.

Christensen, D. 1996. "Dutch Book Arguments Depragmatized." *Journal of Philosophy* 93:450–79.

Coase, R. 1960. "The Problem of Social Costs." *Journal of Law and Economics* 3:1–44.

Coleman, J., and C. Morris. 1998. *Rational Commitment and Social Justice*. Cambridge: Cambridge University Press.

Cranor, C. 1993. *Regulating Toxic Substances*. New York: Oxford University Press.

Crimmins, M. 1992. *Talk about Beliefs*. Cambridge, MA: MIT Press.

D'Aspremont, C., and L. Gevers. 1977. "Equity and the Informational Basis of Collective Choice." *Review of Economic Studies* 44:199–209.

Davidson, D. 1980. *Essays on Actions and Events*. Oxford: Oxford University Press.

1984. *Inquiries into Truth and Interpretation*. Oxford: Oxford University Press.

Davis, W. 1982. "Weirich on Conditional and Expected Utility." *Journal of Philosophy* 79:342–50.

Dayton, E. 1979. "Toward a Credible Act-Utilitarianism." *American Philosophical Quarterly* 16:61–6.

Dworkin, R. 1978. *Taking Rights Seriously*. Cambridge, MA: Harvard University Press.

Eells, E. 1982. *Rational Decision and Causation*. Cambridge: Cambridge University Press.

Eells, E., and W. Harper. 1991. "Ratifiability, Game Theory, and the Principle of Independence of Irrelevant Alternatives." *Australasian Journal of Philosophy* 69:1–19.

Ellsberg, D. 1961. "Risk, Ambiguity, and the Savage Axioms." *Quarterly Journal of Economics* 75:643–69.

Elster, J. 1989. *The Cement of Society.* Cambridge: Cambridge University Press.

Etzioni, A. 1986. "The Case for a Multiple-Utility Conception." *Economics and Philosophy* 2:159–83.

Feigl, H. 1950. "De Principiis Non Disputandum... ?" In M. Black, ed., *Philosophical Analysis*, pp. 119–56. Ithaca, NY: Cornell University Press.

Feldman, F. 1986. *Doing the Best We Can.* Dordrecht: Reidel.

Fischhoff, B., S. Lichtenstein, P. Slovic, S. Derby, and R. Keeney. 1981. *Acceptable Risk.* Cambridge: Cambridge University Press.

Fodor, J. 1998. *In Critical Condition.* Cambridge, MA: MIT Press.

Foley, R. 1993. *Working without a Net.* New York: Oxford University Press.

Franklin, B. 1945. *Benjamin Franklin's Autobiographical Writings.* Ed. C. van Doren. New York: Viking Press.

Fuchs, A. 1985. "Rationality and Future Desires." *Australasian Journal of Philosophy* 63:479–84.

Gärdenfors, P. 1988. *Knowledge in Flux.* Cambridge, MA: MIT Press.

Gärdenfors, P., and N. Sahlin. 1982. "Unreliable Probabilities, Risk Taking, and Decision Making." *Synthese* 53:361–86.

Garrett, D., and E. Barbanell, eds. 1997. *Encyclopedia of Empiricism.* Westport, CT: Greenwood Press.

Gauthier, D. 1986. *Morals by Agreement.* Oxford: Oxford University Press.

Gibbard, A., and W. Harper. 1981. "Counterfactuals and Two Kinds of Expected Utility." In W. Harper, R. Stalnaker, and G. Pearce, eds., *Ifs*, pp. 153–90. Dordrecht: Reidel.

Gilbert, M. 1996. *Living Together.* Lanham, MD: Rowman and Littlefield.

Goldman, A. 1972. "Toward a Theory of Social Power." *Philosophical Studies* 23:221–68.

Goldman, A. I. 1995. "Simulation and Interpersonal Utility." *Ethics* 105:709–26.
   1996. "The Science of Consciousness and the Publicity of Science." Paper presented at the Central States Philosophical Association meeting, Kansas City, October 11.
   1997. "Science, Publicity, and Consciousness." *Philosophy of Science* 64:525–45.

Good, I. J. 1967. "On the Principle of Total Evidence." *British Journal for the Philosophy of Science* 17:319–21.

Hahn, R., ed. 1996. *Risks, Costs, and Lives Saved.* New York: Oxford University Press.

Hamminga, B., and N. De Marchi. 1994. *Idealization VI: Idealization in Economics.* Amsterdam: Rodopi.

Hammond, P. 1976. "Equity, Arrow's Conditions, and Rawls' Difference Principle." *Econometrica* 44:793–804.

Hampton, J. 1994. "The Failure of Expected-Utility Theory as a Theory of Rationality." *Economics and Philosophy* 10:195–242.

Harsanyi, J. 1955. "Cardinal Welfare, Individual Ethics, and Interpersonal Comparisons of Utility." *Journal of Political Economy* 63:309–21.

1967–8. "Games of Incomplete Information Played by Bayesian Players, Parts I–III." *Management Science* 14:159–82, 320–34, 486–502.

1977. *Rational Behavior and Bargaining Equilibrium in Games and Social Situations.* Cambridge: Cambridge University Press.

Hempel, C. 1965. *Aspects of Scientific Explanation.* New York: Free Press.

1966. *Philosophy of Natural Science.* Englewood Cliffs, NJ: Prentice-Hall.

Hölder, O. 1901. "Die Axiome der Quantität und die Lehre vom Mass." *Berichte der Sächsisschen Gesellschaft der Wissenschaften, math. phys. Klasse* 53: 1–64.

Horgan, T. 2000. "The Two-Envelope Paradox, Nonstandard Expected Utility, and the Intensionality of Probability." *Noûs* 34:578–603.

Horwich, P. 1998. *Meaning.* Oxford: Oxford University Press.

Hylland, A., and R. Zeckhauser. 1979. "The Impossibility of Bayesian Group Decisions with Separate Aggregation of Beliefs and Values." *Econometrica* 47:1321–36.

Jackson, F. 1986. "What Mary Didn't Know." *Journal of Philosophy* 83:291–5.

ed. 1991. *Conditionals.* Oxford: Oxford University Press.

Jeffrey, R. 1983. *The Logic of Decision.* 2d edition. Chicago: Chicago University Press.

1992. *Probability and the Art of Judgment.* Cambridge: Cambridge University Press.

Joyce, J. 1998. "A Nonpragmatic Vindication of Probabilism." *Philosophy of Science* 65:575–603.

1999. *The Foundations of Causal Decision Theory.* Cambridge: Cambridge University Press.

Kagan, S. 1988. "The Additive Fallacy." *Ethics* 99:5–31.

Kahneman, D., and A. Tversky. 1979. "Prospect Theory." *Econometrica* 47:263–91.

Kalai, E., and M. Smorodinsky. 1975. "Other Solutions to Nash's Bargaining Problem." *Econometrica* 43:513–18.

Kaplan, D. 1969. "Quantifying In." In D. Davidson and J. Hintikka, eds., *Words and Objections: Essays on the Work of W. V. Quine.* New York: Humanities Press.

1989. "Demonstratives." In J. Almog, J. Perry, and H. Wettstein, eds., *Themes from David Kaplan,* pp. 481–614. Oxford: Oxford University Press.

Kaplan, M. 1996. *Decision Theory as Philosophy.* Cambridge: Cambridge University Press.

Kavka, G. 1983. "The Toxin Puzzle." *Analysis* 43:33–6.

Keeney, R., and H. Raiffa. 1976. *Decisions with Multiple Objectives.* New York: Wiley.

Kleindorfer, P., H. Kunreuther, and P. Schoemaker. 1993. *Decision Sciences.* Cambridge: Cambridge University Press.

Korsgaard, C. 1983. "Two Distinctions in Goodness." *Philosophical Review* 92:169–95.

Krantz, D., R. Luce, P. Suppes, and A. Tversky. 1971. *Foundations of Measurement.* Vol. 1. New York: Academic Press.

Kripke, S. 1980. *Naming and Necessity.* Cambridge, MA: Harvard University Press.

Kyburg, H. 1974. *The Logical Foundations of Statistical Inference.* Dordrecht: Reidel.

Lehrer, K., and C. Wagner. 1981. *Rational Consensus in Science and Society.* Dordrecht: Reidel.

Lemos, N. 1994. *Intrinsic Value.* Cambridge: Cambridge University Press.

Levi, I. 1980. *The Enterprise of Knowledge.* Cambridge, MA: MIT Press.

——— 1996. *For the Sake of Argument.* Cambridge: Cambridge University Press.

——— 1997. *The Covenant of Reason.* Cambridge: Cambridge University Press.

Lewis, D. 1976. "Probabilities of Conditionals and Conditional Probabilities." *Philosophical Review* 85:297–315.

——— 1981. "Causal Decision Theory." *Australasian Journal of Philosophy* 59:5–30.

——— 1983. *Philosophical Papers.* Vol. 1. New York: Oxford University Press.

——— 1986a. "Probabilities of Conditionals and Conditional Probabilities II." *Philosophical Review* 95:581–9.

——— 1986b. *Philosophical Papers.* Vol. 2. New York: Oxford University Press.

——— 1998. *Papers in Philosophical Logic.* Cambridge: Cambridge University Press.

Libby, R., and P. Fishburn. 1977. "Behavioral Models of Risk Taking in Business Decisions: A Survey and Evaluation." *Journal of Accounting Research* 15:272–92.

Lind, H. 1993. "A Note on Fundamental Theory and Idealizations in Economics and Physics." *British Journal for the Philosophy of Science* 44:493–503.

Lopes, L. 1988. "Economics as Psychology: A Cognitive Assay of the French and American Schools of Risk Theory." In Bertrand Munier, ed., *Risk, Decision and Rationality*, pp. 405–16. Dordrecht: Reidel.

Luce, R. D., and H. Raiffa. 1957. *Games and Decisions.* New York: Wiley.

Machina, M. 1982. "'Expected Utility' Analysis without the Independence Axiom." *Econometrica* 50:277–323.

MacKay, A. 1980. *Arrow's Theorem: The Paradox of Social Choice.* New Haven: Yale University Press.

Maher, P. 1993. *Betting on Theories.* Cambridge: Cambridge University Press.

Markowitz, H. 1959. *Portfolio Selection.* New York: John Wiley & Sons.

Mayo, D., and R. Hollander, eds. 1991. *Acceptable Evidence.* New York: Oxford University Press.

McClennen, E. 1990. *Rationality and Dynamic Choice.* Cambridge: Cambridge University Press.

——— 1992. "Rational Choice in the Context of Ideal Games." In C. Bicchieri and M. Chiara, eds., *Knowledge, Belief, and Strategic Interaction*, pp. 47–60. Cambridge: Cambridge University Press.

Mullen, J., and B. Roth. 1991. *Decision Making.* Savage, MD: Rowman and Littlefield.

Nash, J. 1950. "The Bargaining Problem." *Econometrica* 18:155–62.

Nozick, R. 1969. "Newcomb's Problem and Two Principles of Choice." In N. Rescher, ed., *Essays in Honor of C. G. Hempel*, pp. 114–46. Dordrecht: Reidel.

——— 1993. *The Nature of Rationality.* Princeton, NJ: Princeton University Press.

Parfit, D. 1984. *Reasons and Persons.* Oxford: Oxford University Press.
Pratt, J. 1964. "Risk Aversion in the Small and in the Large." *Econometrica* 32:122–36.
Pratt, J., and R. Zeckhauser, eds. 1985. *Principals and Agents.* Boston: Harvard Business School Press.
Putnam, H. 1975. "The Meaning of Meaning." In *Mind, Language, and Reality: Philosophical Papers,* 2:215–71. Cambridge, MA: MIT Press.
Quinn, W. 1974. "Theories of Intrinsic Value." *American Philosophical Quarterly* 11:123–32.
Rabinowicz, W. 1995. "To Have One's Cake and Eat It, Too." *Journal of Philosophy* 92:586–620.
Raiffa, H. 1968. *Decision Analysis.* Reading, MA: Addison-Wesley.
Ramsey, F. 1931. *The Foundations of Mathematics.* Ed. R. Braithwaite. New York: Harcourt.
Rawling, P. 1993. "Choice and Conditional Expected Utility." *Synthese* 94:303–28.
   1997. "Expected Utility, Ordering, and Context Freedom." *Economics and Philosophy* 13:79–86.
Resnik, M. 1987. *Choices.* Minneapolis: University of Minnesota Press.
Richard, M. 1990. *Propositional Attitudes.* Cambridge: Cambridge University Press.
Roberts, K. 1980. "Interpersonal Comparability and Social Choice Theory." *Review of Economic Studies* 67:421–39.
Robinson, A. 1966. *Non-Standard Analysis.* Amsterdam: North Holland.
Roth, A. 1979. *Axiomatic Models of Bargaining.* Berlin: Springer-Verlag.
   ed. 1985. *Game-Theoretic Models of Bargaining.* Cambridge: Cambridge University Press.
Rubinstein, A. 1982. "Perfect Equilibrium in a Bargaining Model." *Econometrica* 50:97–109.
   1998. *Modeling Bounded Rationality.* Cambridge, MA: MIT Press.
Salmon, N., and S. Soames, eds. 1988. *Propositions and Attitudes.* New York: Oxford University Press.
Savage, L. 1954. *The Foundations of Statistics.* New York: Wiley.
Schick, F. 1991. *Understanding Action.* Cambridge: Cambridge University Press.
   1997. *Making Choices.* Cambridge: Cambridge University Press.
Scriven, M. 1959. "Explanation and Prediction in Evolutionary Theory." *Science* 130:477–82.
Seidenfeld, T., J. Kadane, and M. Schervish. 1989. "On the Shared Preferences of Two Bayesian Decision Makers." *Journal of Philosophy* 86:225–44.
Sen, A. 1970. *Collective Choice and Social Welfare.* San Francisco: Holden-Day.
   1985. "Rationality and Uncertainty." *Theory and Decision* 18:109–27.
Shapley, L. 1969. "Utility Comparisons and the Theory of Games." In G. Guilbaud, ed., *La Decision,* pp. 251–63. Paris: CNRS.
Shubik, M. 1982. *Game Theory in the Social Sciences.* Cambridge, MA: MIT Press.
Skyrms, B. 1980. *Causal Necessity.* New Haven: Yale University Press.
   1984. *Pragmatics and Empiricism.* New Haven: Yale University Press.
   1985. "Discussion: Ultimate and Proximate Consequences in Causal Decision Theory." *Philosophy of Science* 52:608–11.

1996. *Evolution of the Social Contract*. Cambridge: Cambridge University Press.

Slote, M. 1989. *Beyond Optimizing*. Cambridge, MA: Harvard University Press.

Sobel, J. 1970. "Utilitarianisms: Simple and General." *Inquiry* 13:394–449.

1976. "Utilitarianism and Past and Future Mistakes." *Noûs* 10:195–219.

1994. *Taking Chances: Essays on Rational Choice*. Cambridge: Cambridge University Press.

1996. "Pascalian Wagers." *Synthese* 108:11–61.

1997. "Cyclical Preferences and World Bayesianism." *Philosophy of Science* 64:42–73.

1998. "Ramsey's Foundations Extended to Desirabilities." *Theory and Decision* 44:231–78.

Stalnaker, R. 1981a. "A Theory of Conditionals." In W. Harper, R. Stalnaker, and G. Pearce, eds., *Ifs*, pp. 41–55. Dordrecht: Reidel.

1981b. "Letter to David Lewis, May 21, 1972." In W. Harper, R. Stalnaker, and G. Pearce, eds., *Ifs*, pp. 151–2. Dordrecht: Reidel.

1981c. "Indicative Conditionals." In W. Harper, R. Stalnaker, and G. Pearce, eds., *Ifs*, pp. 193–210. Dordrecht: Reidel.

1984. *Inquiry*. Cambridge, MA: MIT Press.

Strasnick, S. 1975. "Preference Priority and the Maximization of Social Welfare." Ph.D. dissertation, Harvard University.

Suzumura, K. 1983. *Rational Choice, Collective Decisions and Social Welfare*. Cambridge: Cambridge University Press.

United States Code. 1995. 1994 edition. Vol. 16. Washington, DC: United States Government Printing Office.

Vallentyne, P. 1987. "Utilitarianism and the Outcomes of Actions." *Pacific Philosophical Quarterly* 68:57–70.

1988. "Teleology, Consequentialism, and the Past." *Journal of Value Inquiry* 22:89–101.

1991. "The Problem of Unauthorized Welfare." *Noûs* 25:295–321.

Vallentyne, P., and S. Kagan. 1997. "Infinite Value and Finitely Additive Value Theory." *Journal of Philosophy* 94:5–26.

Van Fraassen, B. 1976. "Probabilities of Conditionals." In W. Harper and C. Hooker, eds., *Foundations and Philosophy of Epistemic Applications of Probability Theory*, vol. 1 of *Foundations of Probability Theory, Statistical Inference, and Statistical Theories of Science*, pp. 261–300. Dordrecht: Reidel.

1983. "Calibration: A Frequency Justification for Personal Probability." In R. Cohen and L. Laudan, eds., *Physics, Philosophy, and Psychoanalysis*, pp. 295–319. Dordrecht: Reidel.

Velleman, D. 1997. "Deciding How to Decide." In G. Cullity and B. Gaut, eds., *Ethics and Practical Reason*, pp. 29–52. Oxford: Oxford University Press.

Vickers, J. 1995. "Value and Probability in Theories of Preference." *Pacific Philosophical Quarterly* 76:168–82.

Vineberg, S. 1997. "Dutch Books, Dutch Strategies and What They Show about Rationality." *Philosophical Studies* 86:185–201.

Viscusi, W. 1998. *Rational Risk Policy*. Oxford: Oxford University Press.

Von Neumann, J., and O. Morgenstern. 1944. *Theory of Games and Economic Behavior.* Princeton, NJ: Princeton University Press.

Weber, M. 1998. "The Resilience of the Allais Paradox." *Ethics* 109:94–118.

Weirich, P. 1977. *Probability and Utility for Decision Theory.* Ph.D. dissertation, University of California, Los Angeles. Ann Arbor, MI: University Microfilms.

———. 1980. "Conditional Utility and Its Place in Decision Theory." *Journal of Philosophy* 77:702–15.

———. 1981. "A Bias of Rationality." *Australasian Journal of Philosophy* 59:31–7.

———. 1982. "Decisions When Desires Are Uncertain." In M. Bradie and K. Sayre, eds., *Reason and Decision*, pp. 69–75. Bowling Green, IN: Bowling Green State University.

———. 1983a. "Conditional Probabilities and Probabilities Given Knowledge of a Condition." *Philosophy of Science* 50:82–95.

———. 1983b. "A Decision Maker's Options." *Philosophical Studies* 44:175–86.

———. 1984a. "Interpersonal Utility in Principles of Social Choice." *Erkenntnis* 21:295–317.

———. 1984b. "The St. Petersburg Gamble and Risk." *Theory and Decision* 17:193–202.

———. 1985. "Probabilities of Conditionals in Decision Theory." *Pacific Philosophical Quarterly* 65:59–73.

———. 1986. "Expected Utility and Risk." *British Journal for the Philosophy of Science* 37:419–42.

———. 1987. "Mean-Risk Decision Analysis." *Theory and Decision* 23:89–111.

———. 1988a. "A Game-Theoretic Comparison of the Utilitarian and Maximin Rules of Social Choice." *Erkenntnis* 28:117–33.

———. 1988b. "Trustee Decisions in Investment and Finance." *Journal of Business Ethics* 7:73–80.

———. 1990. "L'utilité collective." In E. Archambault and O. Arkhipoff, eds., *La comptabilité nationale face au défi international*, pp. 411–22. Paris: Economica.

———. 1991a. "Group Decisions and Decisions for a Group." In A. Chikán, ed., *Progress in Decision, Utility and Risk Theory*, pp. 271–9. Norwell, MA: Kluwer.

———. 1991b. "The General Welfare as a Constitutional Goal." In D. Speak and C. Peden, eds., *The American Constitutional Experiment*, pp. 411–32. Lewiston, NY: Edwin Mellen Press.

———. 1993. "Contractarianism and Bargaining Theory." In John Heil, ed., *Rationality, Morality, and Self-Interest*, pp. 161–74. London: Rowman and Littlefield.

———. 1998. *Equilibrium and Rationality: Game Theory Revised by Decision Rules.* Cambridge: Cambridge University Press.

———. 2001. "Risk's Place in Decision Rules." *Synthese* 126:427–41.

———. Forthcoming. "Belief and Acceptance." In I. Niiniluoto, M. Sintonen, and J. Wolenski, eds., *Handbook of Epistemology*. Dordrecht: Kluwer.

Index

consistency of utility analyses, 164–5, 244

contextualism, 27–9, 206; advantages of, 31–3; *see also* operationism

coordination, 176–7; through agreement, 177

core of a cooperative game, 209

Cranor, C., 233 n7

Crimmins, M., 44 n3, 93

Dasgupta, P., 206

D'Aspremont, C., 205

Davidson, D., 30, 104

Davis, W., 143 n23

Dayton, E., 83–4

decision, 79; and act, 85; consequences of, 79–80; contents of, 79; costs of, 79; of a group, 174; inefficacious, 82; null, 82; paternalistic, 216; proxy, 217; trustee, 217

decision space, 4, 241

decision theory: American school of, 39–40 n34; causal, 116, 131; contextualist, 112; empirical, 36–7; evidential, 131; French school of, 39–40 n34; normative, 1; philosophical, 36–7

degree of belief, 101

degree of desire, 101

De Marchi, N., 22 n12

desire: basic intrinsic, 42, 48; basic intrinsic and of a group, 196; conditional intrinsic, 55; extrinsic, 43, 53, 66; final, 43; of a group, 173; intrinsic, 42; partial, 157; second-order intrinsic, 53

dimension (for utility analysis), 4; basic, 10, 150; causal, 7, 91–2; derivative, 7, 11; of goals, 6; of people, 6; of possible worlds, 6; spatial, 7, 242; spatiotemporal, 7; temporal, 7, 91–2, 242

direct inference, 189–90 n5

direct reference, 44, 81

Dworkin, R., 157

Eells, E., 125 n8, 126, 131, 180–1 n5

Ellsberg's paradox, 118–23

Elster, J., 206

empiricism, 27–9

equilibrium: among desires, 186 n10; and the intersection point for

bargaining, 214–15; optimal, 177

Etzioni, A., 8

expected desirability principle, 103

explanation, 33–8; causal, 34–5; partial, 20, 35–6

Feigl, H., 18

Feldman, F., 43 n1, 69 n13

Fischhoff, B., 233 n7

Fishburn, P., 227

Fodor, J., 27 n18

Foley, R., 20

Franklin, B., 12 n7

Freeman, R., 218 n1

Frege, G., 44 n2, 92–3

Fuchs, A., 23 n14

games: cooperative, 204; noncooperative, 211–15

Gärdenfors, P., 118 n2, 134 n18

Garrett, D., 28 n22

Gauthier, D., 17 n10, 206

Gevers, L., 205

Gibbard, A., 92 n8, 126, 130–3

Gilbert, M., 177 n4

goals, 41; of rational decision, 83–4; stability of, 55

Goldman, A., 183 n8

Goldman, A. I., 38 n33, 182 n6

Good, I. J., 193 n15

group, 176; as an agent, 176; preferences of, 179–81, 205–6; choice function for, 179

guidance, 22

Hagen, O., 39 n34

Hahn, R., 233 n7

Hamminga, B., 22 n12

Hammond, P., 205

Hampton, J., 111 n22

Hardcastle, G., 28 n22

Harper, W., 92 n8, 126, 130–3, 180–1 n5

Harsanyi, J., 189 n12, 191, 203 n24, 206

Harsanyi Doctrine, 189 n12

Harsanyi's theorem, 191–2

Hempel, C., 29 n25, 34 n29

Hölder, O., 108

Hollander, R., 233 n7

Horgan, T., 94 n11

269

271

utilitarianism, 186
utility, 61–2, 100–5; accessibility of, 98,
101; canonical method of computing,
245–50; causal, 14, 91;
comprehensive, 41, 62, 91;
conditional, 75, 98, 135–46, 160;
conditional group, 187; direct, 62 n9;
doubly conditional, 140–6; expected,
75; expected conditional, 146;
formation of, 113; group, 14, 29,
173–4, 179, 185; group intrinsic, 197;
indirect, 62 n9; interpersonal, 29,
181–2, 205–6, 208; intrinsic, 41, 47,
61–6; measurement of, 110; partial,
156, 166, 225; report of, 93–4; social,
186; subjective, 100; temporal, 14,
91–2
utility analyses conjointly made, 2–9;
expected and group, 168, 188–96,
231; expected and intrinsic, 9, 150–1,
154–5, 248–9, 253–4, 256; expected,
group, and intrinsic, 232, 240, 256–9;
expected, temporal, and group, 10;
group and intrinsic, 196–7; hybrid,
255–6; partial and expected, 160, 256;
partial and group, 231–2
utility analysis, 1; BIT by BIT, 155–6,
165; expected, 2, 75–7, 105–6, 144–6;
expected conditional, 106–7, 146–8;

fundamental unit for, 248–9; given
certainty, 151–3; given uncertainty,
153–5; group, 2, 168, 230–2; hybrid,
165; intrinsic, 1, 29, 41, 151, 152, 154;
mean-risk, 162; mean-variance, 162;
multiattribute, 8; multidimensional,
1, 241; partial, 159, 166–7, 254–5;
temporal, 164 n7; viable hybrid, 165;
world by world, 155–6, 165